Such Silver Currents

Coloured lithograph portraying vessel Balmoral Castle which carried William and Lucy Clifford to Madeira (1878).

Such Silver Currents

The Story of William and Lucy Clifford
1845–1929

M. Chisholm

Ⓛ
The Lutterworth Press

THE LUTTERWORTH PRESS
P.O. Box 60
Cambridge
CB1 2NT

www.lutterworth.com
publishing@lutterworth.com

Paperback ISBN: 978 0 7188 9567 9
ePDF ISBN: 978 0 7188 4827 9
ePub ISBN: 978 0 7188 4828 6

British Library Cataloguing in Publication Data
A catalogue record is available from the British Library

Copyright © M. Chisholm, 2002

First published in 2002, 2021

All rights reserved. No part of this edition may be reproduced,
stored electronically or in any retrieval system, or transmitted
in any form or by any means, electronic, mechanical,
photocopying, recording, or otherwise, without
prior written permission from the Publisher
(permissions@lutterworth.com).

In memory of my mother

Oh, two such silver currents when they join
Do glorify the banks that bound them in.

Lucy Clifford's epitaph
Highgate Cemetery, London

Contents

Illustrations	vi
Preface and Acknowledgements	vii
Foreword by Sir Michael Atiyah	x
Introduction: Setting the Scene	1

PART I: 1845-1875

Lucy's Early Years	6
William's Childhood in Exeter	13
The Cambridge Years	18
'The Great Scientific Missionary': University College, London	28
Life in London: Friendship with George Eliot and George Lewes	41
William's Death in Madeira	59

PART II: LUCY ALONE, 1879-1929

Beginning Again	69
Friendship with Rudyard Kipling	73
Literary Critic and Hostess	79
Friendship with William and Henry James	85
Transatlantic Friendships: Oliver Wendell Holmes Jr. and J.R. Lowell	102
Bloomsbury Connections: Leslie Stephen and Virginia Woolf	110
London Friends at the Turn of the Century	123

PART III: HERITAGE, LITERARY AND SCIENTIFIC

Lucy Clifford's Books and Plays	134
The Clifford Heritage	159
Afterword by Sir Roger Penrose	178
Bibliography	185
Index	193

Illustrations

Coloured lithograph portraying vessel Balmoral Castle (1878)	ii
Park Place, Exeter, William Clifford's childhood home.	14
W.K. Clifford as a child.	15
W.K. Clifford, 1868.	24
W.K. Clifford with the English Eclipse Expedition, 1870.	26
Page Two of William Kingdon Clifford's *Mathematical Fragments*.	49
William Kingdon Clifford. Painting by the Hon John Collier.	60
The gardens at Miles Carmo Hotel, Funchal, circa 1870.	61
Miles Carmo Hotel. Oil sketch by the Hon John Collier, 1879.	62
Rudyard Kipling. Portrait by the Hon John Collier, 1891.	75
Rudyard Kipling. Pencil sketch by his father.	76
Henry James. Seventieth birthday portrait by John Singer Sargent.	99
Oliver Wendell Holmes.	104
Dogs near the lighthouse at St. Ives, by Leslie Stephen, 1875.	112
Bears by Leslie Stephen.	114
Ethel Dilke. Pencil sketch by Edward Burne Jones.	131
'Mrs Keith'. Pencil sketch by the Hon John Collier.	140
Lucy Clifford. (undated photograph).	152
Cartoon of W.K. Clifford at Cambridge, by Dave Chisholm.	162

Photographs and drawings of Lucy Clifford, Ethel Dilke and William Cliffordas a boy, are included by permission of Alice Dilke and her daughter Caroline. Also from the Dilke collection came the drawings by Leslie Stephen. The reproduction of the Collier portrait of William Kingdon Clifford is included by permission of the President and Council of the Royal Society. Lady Jennings provided the photograph of William Clifford (held in the Peter Ward collection), which is used on the cover of the book. The Royal Astronomical Society Library gave permission to include the photograph of William Clifford and the 1870 Eclipse Expedition team. The John Collier oil sketch of the Miles Carmo Hotel garden is reproduced by permission of Lucilla van den Bogaerde. The National Trust Photographic Library, John Hammond, and the copyright holder Susannah Greenwood, gave permission to use the Collier 1891 portrait of Rudyard Kipling. Lesley Farr of the University of Kent Design Studio helped with the cover design. My thanks go to all of these for their support.

Preface and Acknowledgements

Since William Clifford, the celebrated mathematician, and Lucy Clifford, his novelist wife, had only four brief years together, it has been necessary, in this biographical study, to chart parts of their lives separately. The individual sections of the book are linked by the overlapping involvement of their wide circle of mutual and famous friends who, between them, introduce and weave together some of the philosophical, literary, political and scientific perspectives of Victorian and early Edwardian times. I have included a mainstream review of both William's and Lucy's writings even though the books are not readily available today. Chapter endnotes detail the sources of references and quoted correspondence.

The great good fortune is that the richest source of unpublished material – the thousand or so letters amassed by Lucy Clifford which form the Valehouse Collection – has been cherished and preserved and eventually made available to me by Alice Dilke, whose late husband, Christopher, was Lucy's grandson. Alice Dilke wears Lucy's wedding ring with William's words 'Careful I'll be to comfort thee' inscribed on it. She has, by her ardent support, enabled Lucy and William and their friends to live again through these pages. Caroline Dilke, who now has charge of the Valehouse Collection, has given me her permission to use it. Access to all of W. K. Clifford's papers has generously been given by his great-grandson, Fisher Dilke.

The William and Lucy Clifford Research Group was formed to prepare the many 1995/6 150th anniversary conferences, exhibitions and talks about the Cliffords. It would be impossible to separate out the items that each of the four members of the group has contributed. The mathematical side of the research has been the special territory of Ruth Farwell and Roy Chisholm. They have been professionally involved in Clifford Algebra for over twenty years. Marysa Demoor of the University of Ghent has a special interest in Lucy Clifford's work as a journalist – especially her connection with the *Athenaeum*. The writing of this book and the specific research for it is my own work, but the pool of collaborative research has been dipped into and used by all of us at different times. Each of us has supported the work of the others and I want here to record, with gratitude, my debt to the group.

I also wish to acknowledge the support and help I have had from libraries and librarians holding the collections of letters, microfilms and books used for my research. Most of the Henry James letters to Lucy Clifford are held in the Houghton Collection I also wish to acknowledge the support and help I have had from libraries and librarians holding the collections of letters, microfilms and books used for my research. Most of the Henry James letters to Lucy Clifford are held in the Houghton Collection at Harvard. Collections of letters and other research sources have been found in The British Library; Trinity College Library, Cambridge; The Bodleian Library; University College Library, London; The Harvard Law School Library; The Brotherton Collection at Leeds Public Library; The Richmond-upon-Thames Borough Library; The National Library of Scotland; The Harry Ransom Humanities Research Centre, Austin, Texas; The Huntington Library, San Marino, California; The Walpole Library at The King's School Canterbury; The Ellen Terry Museum; The Library at the University of Kent at Canterbury; The National Portrait Gallery Archive; Colindale Newspaper Archive and the Valehouse Collection. Permission to quote from Henry James's letters to Lucy Clifford has been generously given by Bay James. The Houghton Library in Harvard University have given permission to use the letters held in their collection under the indexes bMS AM 123.716(1), bMS Am 1094(130), bMS Am 1659(44-46), bMS Am 148(178-180), bMS Am 1094(636), and bMSAm 765. The Harvard Law School Library has given permission to quote from letters held in the Oliver Wendell Holmes Papers. A. P. Watt on behalf of The National Trust for Places of Historic Interest or Natural Beauty have given permission to quote from three unpublished letter written by Rudyard Kipling to Lucy Clifford. Susannah Greenwood has most generously given me permission to reproduce the John Collier portrait of Rudyard Kipling and Collier's illustration for *Mrs Keith's Crime.* The Society of Authors, as the Literary Representatives of the Estates of Virginia Woolf and Clive Bell, has given permission to use the unpublished letter from Virginia Woolf to Lucy Clifford held in the Monk's House Papers at the University of Sussex, and the letter from Clive Bell to Frances Partridge. The Executors of the Virginia Woolf Estate and the Hogarth Press have given permission to quote from the *Letters of Virginia Woolf*, edited by Nicolson and Trautman, and *The Diary of Virginia Woolf* edited by A. Oliver Bell. Macmillan Press has allowed me to quote from *The Collected Letters of Rudyard Kipling* edited by Thomas Pinney. Finally, and most importantly, there is my husband Roy. He is a pioneer in the field of Clifford Algebras and a tower of strength and reassurance to me in my endeavour to research and relate the story of the Cliffords. The chapter on *The Clifford Heritage* was written jointly with him.

Foreword
by Sir Michael Atiyah, O.M., F.R.S.

William and Lucy Clifford were a remarkable couple, he a mathematician of genius and an outstanding public speaker, she a literary figure in her own right who was at the centre of a circle of eminent Victorians. Together they spanned the intellectual life of London in its heyday in the 19th century. It is a fascinating story of wide interest and one that fully deserves the loving treatment presented here by Mrs Chisholm. Based largely as it is on the unpublished letters of Lucy Clifford, it is a substantial addition to our understanding of a great period in English history.

William Clifford was a student and a Fellow of Trinity College Cambridge, home of many mathematicians from Isaac Newton onwards. It is therefore entirely appropriate that the College should have assisted in the publication of this book. It is a small tribute to the memory of an exceptional man.

Introduction
Setting the Scene

On 7 April, 1875, mathematics students turning up for their morning lecture at University College London were surprised to see a message chalked on the blackboard. It read, 'I am obliged to be absent on important business which will probably not occur again.' The Professor of Applied Mathematics and Mechanics, William Kingdon Clifford, was getting married on that day. He had come to London from Trinity College, Cambridge, in 1871 and was twenty-nine years old. He was highly intellectual, tremendously popular, slightly eccentric and such a brilliant lecturer that he had taken the academic world by storm. His bride to be was Sophia Lucy Jane Lane. She was one year younger than William Clifford and had already begun to establish herself as a novelist and journalist. He and Lucy made a most attractive and lively couple and they drew around them a wide circle of friends from all walks of life. In 1874, when their engagement had been announced, William Clifford received this letter:

> Few things have given us more pleasure than the intimation in your note that you had a fiancée. May she be the central happiness and motive force of your career, and by satisfying the affections, leave your rare intellect free to work out its glorious destiny. For, if you don't become a glory to your age and time, it will be a sin and a shame. Nature doesn't often send forth such gifted sons, and when she does, Society usually cripples them. Nothing but marriage – a happy marriage – has seemed to Mrs Lewes and myself wanting to your future.[1]

The letter was from his close friend, the publisher, writer, and philosopher George Henry Lewes. Mrs Lewes was, of course, George Eliot. Her many books – among them, *Adam Bede*, *The Mill on the Floss*, *Silas Marner* and *Middlemarch* – had made her famous. George Lewes, although already married and with children, had nevertheless formed an association with her and they lived together as 'Mr and Mrs Lewes' They both loved William Clifford, and he and Lucy were frequent visitors at *The Priory*, their London home. George Eliot's letter, in some ways foresaw the pattern of William and Lucy's life together. Lucy did become 'central to his happiness', William did become a 'glory to his age and time', and his 'glorious destiny' has been achieved in spite of his early death.

The nineteenth century was a time of enormously exciting scientific discovery and progress, and William Clifford, who was judged in his lifetime to be one of the most outstanding scholars of the century, made important contributions to that progress. His books, papers, and brilliant researches, many of which were left incomplete at his premature death, were assembled and published later by his colleagues. Even the greatest scientific minds of that time, however, could never have predicted that today, in less time than it takes to read the first page of this book, regular electronic bulletins would be flashed around the world linking the groups of mathematicians and physicists whose researches are associated in some way with this nineteenth-century scholar.

Bertrand Russell noted in 1945, that, besides being a mathematician, Clifford was a great philosopher who: 'saw all knowledge, even the most abstract, as part of the general life of mankind, and as concerned in the endeavour to make human existence less petty, less superstitious and less miserable'.[2]

Russell also wrote of Clifford's book, *Common Sense of the Exact Sciences*, 'All that is said in it on the relation of geometry to physics is entirely in harmony with Einstein's theory of gravitation, which was published thirty-six years after Clifford's death.'[3]

Among William Clifford's many friends were Thomas Huxley, George Eliot, Lord Morley, Frederick Pollock, John Tyndall, and Leslie Stephen. Among his academic contemporaries were James Clerk Maxwell, Lord Kelvin, Arthur Cayley, J.J. Sylvester, Oliver Lodge, Karl Pearson, who succeeded him at University College, and H.J.S. Smith of Oxford. In Europe there were C.F. Klein, H. von Helmholtz, Hermann Grassmann, and Ludwig Boltzmann, and in America, J. Willard Gibbs and C.S. Pierce.

A biographical study of William Clifford was made in 1879, the year of his death, by Frederick Pollock. His lovingly written introduction to William's *Lectures and Essays*[4] provided a contemporary picture of this mathematical genius. There is no previous biography of Lucy Clifford and piecing together the early part of her life presented some problems.

Biographers, rustling through the dry leaves of the past and rearranging them in order to reveal their story, sometimes uncover forgotten events and sometimes, too, uncover events that were intended to remain hidden. So it was with research into the life of Sophia Lucy Jane Lane. She was born in 1846. After the early death of her husband she became a successful writer of books and plays. She formed important friendships and she left some unsolved mysteries behind her. Three of the most famous American men of the age were extremely fond of her, and one of them, Henry James, was a very close friend indeed. He wrote to her as his 'Beloved girl' and singled her out with a bequest in his will. In the Valehouse Collection,[5] there are only ten letters from Henry James to Lucy and nine of them are undated. This small number is surprising until we realise that it is actually an indication of how generous Lucy was. She had saved over seventy of James's precious letters and she might well have hung on to them. However, when she heard that Percy Lubbock was preparing an edition of the James letters, she packed up sixty-four of the most interesting ones and sent them to Theodora Bosanquet, Henry James's amanuensis. She typed them out for Lubbock who subsequently published eight of them in full. Leon Edel drew upon them for his *Life of Henry James* and published four of them in his editions of James's letters. Percy Lubbock promised to return the originals to Lucy after they had been copied but no trace of them has been found. However, the typed copies are held at the Houghton Library in Harvard and the full collection, edited and annotated by Marysa Demoor and myself, is published in the *English Literary Studies* 1999 series from the University of Victoria under the title *'Bravest of women and finest of friends': Henry James's Letters to Lucy Clifford*. In one of these lovely letters to her he wrote, 'But what a life you lead! I feel – beside you – like a slug in a damp village garden: you the gorgeous butterfly in the social (and other) blue.'[6] One of the clearest indications of Henry James's feeling for her was shown in the summer of 1912 when he was sixty-nine and Lucy was sixty-six. He must have disappointed Lucy in some way and she let him know of it. He reassured her with these words:

> Dearest Lucy,
> What shall I say? When I love you so very, very much and see you nine times for the once that I see Others! Therefore I think that – if you want it made plain to the meanest intelligence – I love you more than I love Others. I am no great protester now, in my somewhat stricken old age; but I am always your devoted old Nevvy and (at 11.15 p.m.) rather fatigued sleepyhead,
> Henry James.[7]

These direct words, uncluttered by the flowery or oblique style typical of so much of Henry James's letter writings, would have left Lucy, and leave us, in no doubt about the place she held in his affections. At the other extreme Virginia Woolf, in her letters and diaries, wrote extraordinarily cruel and disparaging words about her.

Lucy lived for fifty years as William Clifford's widow. Her writing, her travels and the care of her two daughters kept her fully occupied. Her many friendships sustained her, and her enduring interest in life kept her youthful. Over the years she had collected together a trunkful of treasured correspondence and in the last months of her life she sorted through that collection and destroyed many letters that she did not want others to read. The rest of the letters – there are nearly one thousand – she replaced in the trunk with a note attached. It read, 'Save these, they will be valuable one day.' These letters – *The Valehouse Collection* – illuminate the story that is to follow.

Lucy Clifford's correspondents came from every sphere of life. She exchanged letters with publishers and prime ministers, ambassadors and actors, scientists, philosophers, writers, poets, and politicians. To the eminent friends she had shared with her husband she added Bernard Shaw, Edmund Gosse, Mrs Humphry Ward, Rudyard Kipling, Sir Sidney Colvin, Arnold Bennett, H.G. Wells, Lord Fitzmaurice, Charles Morgan, Ellen Terry, Frederic Harrison, and many, many more. The distinguished Americans, James Russell Lowell and Oliver Wendell Holmes Jr., were two of her very special friends, and Dr Page, the American Ambassador to London during the years of the war, was also a frequent visitor to her home. One of her daughters married

Fisher Wentworth Dilke, eldest nephew of Sir Charles Dilke the politician who in 1885, as a result of what was one of the most publicised political sex-scandals ever, lost his chance of following Gladstone as Prime Minister.[8]

Lucy Clifford's novels and plays were popular when she wrote them and were set in the rapidly changing times in which she lived. Through them she illustrated the restricting formalities of the age. When she was a young woman, marriage could lead to subjugation, and divorce could lead to ostracism for the women concerned. By the time of her death in 1929, social conventions, especially those regarding the role of women in society, were relaxing fast. The First World War and the Women's Suffrage Movement had swept aside many of the confinements and restraints that had so frustrated the lives of Victorian and early Edwardian women. Many women writers were casualties of the changes in conventional thinking that followed those events. New-style readers wanted new-style ideas, and Lucy suffered the fate of many other women writers from the 1890s and turn-of-the-century – she and her books are little known today.

Lucy was at the centre of London society and supremely well placed to write a fascinating autobiography, and yet she never produced one. She told Virginia Woolf that she would never do it, for 'Lucy Clifford never gives away a secret'.[9] That may well have been her reason, because loyalty to friends was one of her strongest tenets, and one of the reasons for her popularity. However, there is another possible explanation. If she had written her life story, she would have had to reveal the truth concerning two small mysteries about her that remained unexplained at the time of her death.

William Clifford's genius was recognised in his lifetime but he died at the age of thirty-three. It is in recent decades and particularly since the first full international Clifford Algebra conference held at the University of Kent at Canterbury in 1985, that the development and application of his mathematical theories have become widespread. Today, all over the world, there are research centres actively extending ideas originating from William Clifford's work. During 1995, the 150th anniversary of his birth, international meetings were held in Canterbury, Mexico, Canada, Madeira – where he died – and at the Newton Institute in Cambridge.

Thomas Hardy, who knew both the Cliffords, wrote that, 'experience is as to intensity and not as to duration'. Lucy was married to William for four brief years and Hardy's words could epitomise their relationship. Her husband remained fresh in her memory and heart throughout her fifty years of widowhood. He is buried in Highgate Cemetery. On his tombstone, which can be found quite close to that of Karl Marx, are the lines he composed himself as he lay dying. They are well known to his followers:

William Kingdon Clifford
Born May 4th, 1845
Died March 3rd, 1879

I was not, and was conceived:
I loved and did a little work.
I am not, and grieve not.

When Lucy died she was buried beside him. These lovely words were added to the tombstone and from them has come the title for this book:

> And
> Lucy, his wife
> Died April 21st, 1929
>
> Oh, two such silver currents when they join
> Do glorify the banks that bound them in.

Notes

1. G.S. Haight (ed.), *The George Eliot Letters*, Oxford University Press, 1954, Vol.VI, p. 102.
2. W.K. Clifford, *The Common Sense of the Exact Sciences*, New Edition 1946, Alfred A. Knopf, New York, p. ix.
3. ibid.
4. W.K. Clifford, *Lectures and Essays*, Macmillan and Co. Ltd. 1879, Introduction.
5. The Valehouse Collection. A private collection of William and Lucy Clifford's letters and papers.
6. M. Demoor and M. Chisholm, (eds.) '*Bravest of women and finest of friends*': *Henry James's Letters to Lucy Clifford*, University of Victoria 1999, item 15.
7. As note 6, item 45.
8. David Nicholls, *The Lost Prime Minister: A Life of Sir Charles Dilke,* Hambledon Press, 1995; R. Jenkins, *Sir Charles Dilke: A Victorian Tragedy*, Collins, 1958.
9. A.O. Bell (ed.), *The Diary of Virginia Woolf,* Hogarth Press, 1977-1984, Vol 2, entry for 24 January 1920.

Part One
1845-1875

Lucy's Early Years
1846-1875

Early years are important and most biographical studies begin with them. However, throughout her life, Lucy Clifford very effectively concealed the details of her childhood. In seeking to unravel the hidden beginnings of Lucy's story we must pick up the threads at the end of her life.

In *The Times* of 22 April, 1929, an obituary appeared. It was a glowing tribute to a woman who was 'distinguished as a novelist and for many years an honoured figure in literary London, a link with great writers and scientists of the Victorian Age'. The obituary included a tribute to her distinguished husband and a review of her books and plays. The names of many of her famous friends were included. She was remembered as 'one of the last survivors of those who attended the famous Sunday afternoon gatherings to do homage to George Eliot at The Priory'. Her contributions to periodical literature and journalism were noted, including her regular contributions to the *Standard*, at that time enjoying its Golden Age under the classical and scholarly control of the renowned editors Mudford and Jeyes. Lucy Clifford was described as:

> possessing an active, acute mind and social gifts which maintained round her a circle continually reinforced by new writers who were attracted by her genuine kindliness. For young ambition she had a particularly tender interest and welcome, and, till her last illness, she never wearied of giving practical help and encouragement.

Two eminent Americans were mentioned:

> Writers who were more her contemporaries, such as J.R. Lowell, who wrote her a remarkable series of letters and Henry James, who remembered her in his will, were attached friends.

The following day a further tribute was added:

> through her tender interest in aspiring youth she had found the *elixir vitae* which defies old age, and to the very last she was young in spirit because she was loved by the younger generation .... In later years at her little house in Chilworth Street she was nearly always at home to welcome, fortify, and encourage any friends who sought escape in that sanctuary from the loneliness of life ... and for them the world will seem cold and inhospitable without her.[1]

The names of Lord Fitzmaurice, Oliver Wendell Holmes Jr, Augustine Birrell and Rudyard Kipling were added to the previous list of her close friends. She had clearly been a much-loved woman and many other glowing tributes followed. But it is through her friendship with Henry James that she is most remembered today.

Henry James died in London on 28 February, 1916. In the December of the previous year he had suffered a stroke. On New Year's Day, as he lay in his bed at Carlyle Mansions, Sir Edmund Gosse had crept into the room to whisper to his friend the news that he had been awarded the Order of Merit. Henry James, private to the last, made no response or indication that he had heard. He waited until the door had been closed on Gosse before he opened his eyes and asked the maid to 'turn off the light and spare my blushes'. Even in this state of near terminal illness and vulnerability he found it easier to dissemble than to show himself. Henry James was exceedingly well known in literary circles in London. He had many friends and even more acquaintances, but for most of them it was difficult to penetrate his habitual camouflage of carefully chosen, carefully delivered words.

Henry James lingered for two months after his stroke. His will was a straightforward document. The James family received the property and insurance. The now-famous Sargent portrait went, as expected, to the National Gallery. As expected, the servants were provided for, and just three of his friends received legacies of £100 – quite a substantial sum in modern terms. These three friends were Jocelyn Persse, Hugh Walpole and Lucy Clifford.

Jocelyn Persse, then aged forty-three, a nephew of Lady Gregory, the Irish playwright and close friend of W.B. Yeats, had enjoyed an 'exquisite relationship' with Henry James. They had met at Mrs Sitwell's wedding to Sidney Colvin in 1903 when he, Lucy Clifford, Robert Browning and Henry James had been among the small group of guests.[3] Henry James and Jocelyn had been immediately attracted to each other. They became the closest and most devoted of friends and were loyal to each other for the remaining thirteen years of James's life.

The young writer Hugh Walpole, James's 'dearest darlingest little Hugh' then aged thirty-two, was another of his beloved young men. He had been close to Henry James since they met in 1909 when James was 66. Walpole wrote to James as '*Très cher Maître*'. James was supportive and tender to him, but one senses that Walpole placed himself at James's side for a different reason. He frankly confessed that in his early days in London he 'ran to writers as a kitten runs to milk'. However, he came to true friendship with Henry James and his book of reminiscences, is dedicated, 'For Henry James, as he knows, with love.'[4]

And then there was Lucy Clifford. How was it that this novelist and playwright, then the seventy year old widow of a mathematical genius, came to be so remembered by one of the most eminent of American writers? Eminent men were not unusual in Lucy's life. Her marriage to William Kingdon Clifford had brought her into contact with a galaxy of famous personalities. As a young widow in 1879, she wrote to her friend Fred (Sir Frederick) Pollock, later Professor of Jurisprudence at Oxford, 'it has been my strange good fortune to know the best and greatest of men'. But to Lucy the greatest of all of them was always her husband and his philosophy of life is reflected in every one of her books and plays.

Until recently the main source of information about Lucy Clifford has come from Leon Edel in his incomparable biographical quintet: *The Life of Henry James*. Edel makes many references to Mrs W.K. Clifford, and a section in Volume Five is devoted to her. Since her books are not now widely read or easy to find, Lucy Clifford is often perceived to have been simply a woman friend of Henry James. They were indeed very close friends and Somerset Maugham provides a delightful vignette of Lucy and Henry James together. He writes of an evening that the three of them spent at the Aldwych Theatre when *The Cherry Orchard* was being performed:

> Henry James was perplexed by The Cherry Orchard . . . and in the second interval he set out to explain to us how antagonistic to his French sympathies was this Russian incoherence. Lumbering through his tortuous phrases, he hesitated now and again in a search for the exact word to express his dismay; but Mrs Clifford had a quick and agile mind; she knew the word he was looking for and every time he paused immediately supplied it. This was the last thing he wanted. He was too well-mannered to protest, but an almost imperceptible expression on his face betrayed his irritation and, obstinately refusing the word she offered, he laboriously sought another, and again Mrs Clifford suggested it only to have it again turned down. It was a scene of high comedy.[5]

Lucy demonstrated, throughout twenty-five or more years of friendship, her very unaffected attitude to him. With her he could be relaxed, and it seems likely that Lucy Clifford, of all his wide circle of friends, saw him at his most natural. But she was not quite what she seemed.

During her last illness in April, 1929, her close friends would have come to the house to visit her thinking they were seeing a woman who was seventy-two years old. In fact she was eighty-two. The incorrect age was entered on the certificate of death witnessed by her son-in law Sir Fisher Wentworth Dilke. Even Lucy's own daughters did not seem to know that she had deceived them about her age. It has never been considered a serious fault for a woman to conceal her age, but Lucy started her deception early by claiming to be only twenty-four when she was really twenty-eight. Her birth certificate states that she was born in Great College Street, Camden Town on 2 August, 1846. However, on all of the official documents that have been traced she never once admitted to 1846 and she avoided mentioning her place of birth. This may have been simply a coquettish whim, but there is evidence that her deception was an attempt to conceal some events of her early years.

Amongst Lucy's possessions is an envelope containing a faded photograph of a distinguished looking, bewhiskered gentleman. Written on it, in Lucy's hand, are the words *'Details of my Grandfather John Brandford Lane. Made Chevalier de Malte or Knight of Malta by King Charles X of France'*. Also in the envelope are the names of three properties he owned in Barbados. The photograph is labelled, 'Sir John Brandford Lane. September 28 1829 aged 39.

Parish Broadwater, Worthing, Sussex'. This was the date of his recorded burial. There were many vicissitudes in the history of the Knights of Malta, resulting in a fragmentation of their records. King Charles X was only on the throne from 1824 to 1830, so Brandford's *Chevalier de Malte* honour must have been given when he was between the ages of thirty-three and thirty-eight. Enquiries in Malta, and from the Order of St John in London, reveal no record of John Brandford Lane, but some of the relevant records are incomplete. There are, however, records of John Brandford Lane as a landowner and member of the House of Assembly in Barbados.

Lucy spoke little of this paternal Barbadian grandfather, or of his son, her father, who was also called John Lane. However, she did encourage the notion that she herself had been born in Barbados. It now seems pretty clear that she was never on that island. The likelihood is that the link with Barbados ended with the 1829 death of Brandford Lane in England. Maybe, to Lucy, Barbados seemed a more attractive and romantic birthplace than Camden Town, and there is little harm in that. However, she was concealing other information about her family background.

When she was left a widow trying to establish herself as a freelance writer, she needed all the help that she could get to break into the world of journalism and publishing. She 'collected' interesting people for her Sunday afternoon salon to make sure that she was at the centre of the literary and social round, and also to gather material for her contributions to the 'gossip' section of the *Athenaeum* between 1881 and 1901. Lucy Clifford was always secretive about her family roots. As far as can be discovered she never told anybody that Thomas Gaspey was her grandfather, that his daughter, Louisa Ellen, had married John Lane in 1844, and that she, Lucy, was the oldest of their children. In fact, in all the hundreds of Lucy's letters that have been retrieved, the name of Gaspey does not appear. Now this name does not perhaps readily spring to mind in the literary world today. But in fact Thomas Gaspey was a most successful journalist, poet, writer, and influential part owner of *The Sunday Times*. He had about twenty, solid, historical volumes to his name. He was for many years the senior member of the council of the Literary Fund. One of his sons had a PhD from Heidelberg and was a successful author. The other was also a prolific writer in prose and verse. What could have made her seek to expunge all tangible links with the maternal side of her family – especially when they had achieved what she aspired to – they had made their reputations in the world of letters. The fact that Lucy was sent to live with her grandfather in Shooters Hill suggests that a family break-up of some sort had taken place. But she always somehow managed to slip out of answering questions about her childhood. One perceptive interviewer observed of Mrs W.K. Clifford that, 'One longs to know what she was as a little girl; and that little girl is just the one point on which it is impossible to excite her interest.'[6] Another interviewer observed 'Mrs Clifford is fluent on all subjects except herself.'[7] Apart from gleaning tiny snips of information from her letters about her life before she met and married the young man who was to become so famous, we can only look for clues in her writings.

Lucy's novels and plays, short stories and children's books are fully discussed in a later chapter. It is interesting to note that all her books and plays are set in the time in which she lived, and the dominant themes in them are the problems faced by women of the middle classes. In several of her novels she is concerned with the plight of divorced women and the dilemma of unhappily married women. Many of the books are about women seeking happiness. Many of them clearly draw upon biographical elements and experiences of Lucy's own life, and, in just one of her books, she recounts the tragic story of the childhood and unhappy early life of a young girl. It is possible that this story may contain clues to the circumstances that led Lucy to falsify her age on her marriage certificate and 'lose' some years of her early life.

Lucy wrote *A Flash of Summer* in 1894 and it was published in episodes in the *Illustrated London News*. In the first publication it was a story of tragic circumstances and unfortunate events ending in a young woman's suicide. In later editions the devastating ending was rewritten to allow for the possibility of a happy conclusion. The beginning part of *A Flash of Summer* covers the life of a little girl who, between the ages of six and seventeen, is living with her widower uncle in Shooters Hill on the outskirts of London. He is not unduly unkind to her, but affection and gentleness are absent from their relationship. She feels, and is, unwanted in the household but no details are given of her parents. The little girl, Katherine, is pensive, sensitive and artistic. The uncle's only son dies and Katherine becomes his sole heir. At seventeen her uncle presses her into marriage with his middle-aged, bullying and money-seeking lawyer friend. The inevitably unhappy marriage ends with Katherine running away to travel incognito in Europe, to meet the love of her life, to lose him because she is afraid to tell him the truth about her marriage, and, eventually, to take her own life. The early descriptions of the Shooters Hill area are so accurate that it is easy today to trace the exact paths that 'Katherine' took in her childhood. She wrote in detail of the little girl who,

> every morning of her life from six to seventeen, save on Sundays and the brief holiday periods she came out of the gate . . . turned to her left and went down the hill, past the church on the one side and the inn on the other, past the stuffed-bird shop and the lane that led to Woolwich, and Ordnance Terrace, and the plantation and the Scrubs.

Every detail is so accurate that Lucy Clifford's novel has been used by the Woolwich Antiquarian Society to help trace lost landmarks in the area.

Back in real life, we find, in the Census of 1871, that Lucy Lane was living with her grandfather Thomas Gaspey, the writer-historian, at 4, Ordnance Terrace, Shooters Hill – a house named and described in *A Flash of Summer*. There are many clear parallels between Lucy Clifford and her fictional young heroine. Unfortunately, the 1861 Census records for this area of London are lost. We do not know when Lucy first moved into her grandfather's home. It is not however, fanciful to draw the conclusion that parts of *A Flash of Summer* are autobiographical, and that the reason for Lucy's distancing herself from the Gaspey family might have been that she was in some way made unhappy by her association with them. The 1871 Census document contains the first 'error' in Lucy's recorded age. She is entered as a magazine writer aged twenty-three. In fact she was twenty-five. By the time she was married in April 1875 she had 'lost' four years by giving her age as twenty-four. When the 1881 Census was taken, Lucy was still taking off four years. However, by the 1891 Census she had lost seven years and this had increased to a ten-year disparity by the time of her death. The only other document traced so far is her application to the Literary Fund in 1880. On that she gave her date of birth as 1849 instead of 1846. Of course, records can contain genuine errors, and one can be unintentionally inaccurate about dates, but these inconsistencies are all in one direction, and it is fair to assume that Lucy really did intend to conceal her age. She certainly never took the opportunity, although she had several, to put the record straight. At times the deception led her into a tangled web of evasions – especially over her early publications. Six months before her death she left instructions that her age should not appear on her coffin and should not be given to any enquirer.

She had been a magazine writer and journalist before she met William Clifford. Many of her articles would have been published anonymously, and many of them have not yet been traced. She did publish in the monthly magazine *Quiver,* and her name, Lucy Lane, appears in the list of contributors in the yearly bound edition for 1872. One of these serialised stories, *The Dingy House at Kensington,* was amended and published anonymously in 1882. It was her first book, but she never openly claimed it as her own. Only once, in 1910, in a personal letter written to her publisher George Bentley, does she admit to having written it. She writes of having been proud to see her name on placards advertising the book, 'when I was an infant' – her gentle way of dodging the age issue.[8] It did well for Lucy and for Ward and Downey in the USA, selling over ten thousand copies. The books on which Lucy's reputation as an author rests are discussed later in this book. However even frail early writings give insight to later strengths, and, in *The Dingy House at Kensington,* we do see the first intimations of Lucy's pragmatic understanding of men which becomes a major characteristic of her mature writing. She has her young heroine note that, 'men object to having the weak points in their character played upon' and 'hate nothing more than being made to look small – especially by a woman'.

The aspect of Lucy's deception about her age that is most puzzling is her husband's role in it. William knew her parents and he knew two of her sisters – he must have known that she was in fact only one year younger than himself. If Lucy had reason for 'losing' four years of her life when she got married, did she hide it from him or did he accommodate it? In his letters to her he often called her 'my child', but this might have been simply an affectionate diminutive. We shall never know what passed between them, and William's illness and early death released Lucy from the danger that any duplicity would be discovered by him.

If we accept that there is an autobiographical element in Lucy's writing we will recognise her self-image easily. Her heroines are never conventionally beautiful but have a haunting and attractive remoteness about them. They have a love of the natural world and a reverence for it. They are thoughtful, sensitive women, not afraid to make their own judgements, and Lucy herself must have been a strikingly interesting young woman when, at the age of twenty-eight, but claiming to be twenty-four, she took the public eye in London society as the wife of one of the most eminent and talked-of academics of the age.

Notes
1. *The Times,* Obituary, April 1919, contributed by J.H. Morgan, K.C.
2. H. Montgomery Hyde, *Henry James at Home*, Methuen 1969, p. 281.
3. Leon Edel, *The Life of Henry James,* Rupert Hart-Davis, 1972, Vol 5, p. 188.
4. Hugh Walpole, *The Apple Trees: Four Reminiscences*, Golden Cockerel Press, 1932.
5. Somerset Maugham, *The Vagrant Mood*, Doubleday 1952, p. 209.
6. Mary Angela Dickens, *A Chat with Mrs W.K. Clifford*, Windsor Magazine 1989. Vol 9, p. 483-485.
7. Whitehall Review, *The Whithall Portrait*, December 1893, No. 917.
8. Bodleian Library, ALS (MS: Eng. Lett. E 116).

William's Childhood in Exeter
1845-1863

On one of her many visits to the Great Exhibition of 1851, Queen Victoria paused in front of a five-foot high glass case in the Arts and Crafts section. It contained a model of the West front of Exeter Cathedral. At first glance it appeared to have been carved in ivory. In fact the model had been made out of the pith of rushes which grow in the Exeter Canal, and had been produced, using only a pair of scissors and a pot of glue, by William Kingdon Clifford's grandmother. The model had taken three years to complete and was correct in every detail. Walter Savage Landor, the art critic, made a valuation of it and it was insured for £3,000 – a huge sum for those days. The Queen was most taken with it and she sent a letter to Mrs Fanny Kingdon congratulating her on her masterpiece.[1] Sadly, no trace of the model can be found today, but a smaller example of her beautiful pith work is still held by the Kingdon family.

Before her marriage, Mrs Kingdon had been Mary-Anne Bodley and was related to Sir Thomas Bodley (1545-1613) who was born in Exeter. He was statesman to Queen Elizabeth and part-founder of the Bodleian Library in Oxford, where his portrait still hangs. He was knighted by King James I in 1603. In many ways he was an academic precursor of William Clifford, for, as well as being a lecturer in natural philosophy at Magdalen College, he was public orator and a brilliant linguist. Another descendent was the famous architect George Frederick Bodley, who designed churches and university buildings at Oxford and Cambridge in the nineteenth century. Mary-Anne Bodley's direct forebears had been in business in Exeter as fringe-makers and coach-lace manufacturers. Mary-Anne's daughter, Fanny Kingdon, married William Clifford, a book and print seller in Starcross near Exeter. It was here that their son, William Kingdon, was born on 4 May, 1845, but no records of the event, apart from the birth certificate, has been traced. The family later took over premises at 23 High Street, Exeter. In 1869 William's father sold the business to C & D Eland and there is still a bookshop of that name in the city. William's mathematical talents were obvious even as a child, and it was always said that he inherited his intellectual abilities from his famous model-making grandmother.

The Clifford family home, where William's mother had been born, is at 9 Park Place, a modest Georgian terrace of 'genteel Dwelling-houses' set back in Longbrook Street. It was noted by George Gissing on a visit to the city.

> In a by-way which declines from the main thoroughfare of Exeter, and bears the name of Longbrook Street, is a row of small houses placed

William Clifford's childhood home. The turret was added later, and the house is now divided into flats. The Exeter Civic Society's commemorative plaque is just visible to the left of the bay window.

above long strips of sloping garden. They are old and plain, with no architectural features calling for mention, unless it be the latticed porch which gives the doors an awkward quaintness . . . . The little terrace may be regarded as urban or rural, according to the tastes and occasions

A charming, undated miniature, which is the only known childhood image of William Clifford

of those who dwell there. In one direction, a walk of five minutes will conduct to the middle of the High Street, and in the other it takes scarcely longer to reach the open country.[2]

Today, the house has an Exeter Civic Society commemorative plaque identifying it as Clifford's childhood home. Park Place was just a short walk from *Clifford's Bookshop* in the High Street. William would have had the run of his father's shop and its stockrooms – a goldmine of reading material for a precocious youngster. In 1857 he may well have walked the hundred or so yards from his home in Park Place, to stand on the nearby bridge and watch the very first South Western Railway steam engine from London pass through. His upbringing seems to have been conventional and strongly religious, but there are very few records of his early life. However, one example of William's mathematical precocity has been remembered. The Clifford family were visiting the Great Exhibition of London in 1851 and staying at William's aunt's home.

She noticed that he was very thoughtful while being put to bed one night and asked what he was thinking about. He replied 'Aunt Annie, I don't think you would know.' It later emerged that he was calculating how many sharp edges of the blade of a penknife could be placed round the wheel of a coach. When he gave his answer, together with the size of the wheel, it was found to be 'correct to within a few figures'. He was six years old.[3]

The great tragedy of William's childhood was the death of his mother when he was nine. She was only thirty-five. His father, who suffered from poor health, remarried and had four more children. In 1878, at the age of fifty-eight, he died in Hyères in southern France where he had gone with his daughters, one of whom was delicate. He had been a much-loved and respected citizen and had served the city well as Alderman and Justice of the Peace. *The Exeter Post* carried loyal tributes to him at his death.

William had gained a place at the Exeter Grammar School and was a pupil there for a few months before being sent, from 1856 to 1860, to the Mansion House School, which became known as Mr Templeton's Academy. This was situated in the impressive Georgian Bedford Circus, which was damaged by bombs in 1942 and later demolished by Exeter City Council. The school had an excellent academic record. William would have been one of about a hundred boy pupils, about half of whom were boarders. When he was fifteen William won a Mathematical and Classical Scholarship to King's College London. In 1860 he would certainly have been aware of the famous 'science versus theology' battle which took place at the British Association Meeting, at which Darwin's *Origin of Species*, published the previous year, was discussed. The interest of the general public was engaged when Wilberforce was reported in the press to have asked Huxley, in open debate, whether his ape ancestors were on his grandfather's or grandmother's side.

At King's, from 1860 till 1863, William studied in the department of General Literature and Science. In his first year he won the Junior Mathematical and Junior Classical Scholarships and also the Divinity Prize. In the two succeeding years he again won the Classical and the Mathematical Scholarships and the Inglis Scholarship for English language, as well as an extra prize for the English Essay. It was at this stage, when he had time to read in the college library, that he delighted in solving and posing mathematical problems. On 1 September, 1863, he wrote to the editor of *The Educational Times*:

> My Dear Sir,
>
> I thank you very much indeed for your kindness, and am sorry that I should have given you the trouble of writing to me. I have been on a walking tour in the North of Devon, or I should have written long before. By to-morrow I hope to send you something, and will do what I can to follow out your kind suggestions: but I am a very junior reader myself. I have had it in mind almost from the time I began to fly kites (I have not yet left off) the problem of finding the form of a kite-string under the action of the wind. On a rough trial the other day, the intrinsic equation seemed not very difficult to obtain; if I get any result I will send it to

you tomorrow. I have been trying to construct a second interpretation of mechanical equations, similar to that of tangential co-ordinates, but have failed hitherto. Being a firm believer in the duality of symbols, I should look upon complete failure as a proof that our symbolical system is wrong. You will be amused by my visionary attempt at obtaining a method of inventing problems by the dozen.[4]

William kept up his interest in kite-flying and in 1877 when on holiday with Lucy at a friend's home in Wales, he constructed a kite 'of unusual dimensions, with tail in proportion' with which he hoped to break all previous records of kite-flying. Unfortunately, while they ate lunch, a flock of sheep and their shepherd became entangled in the carefully laid out strings and the experiment had to be abandoned.[5]

He rounded off his scholarly achievements and fulfilled his early promise by gaining a Foundation Scholarship to Trinity College, and went up to Cambridge in 1863. His degree, a BA in Mathematics and Natural Philosophy, was awarded later in 1867. Before he left King's, when he was only eighteen, he produced one of his earliest papers, *The Analogues of Pascal's Theorem*, which was published in *The Quarterly Journal of Pure and Applied Mathematics* in March 1863.

Notes

1. Reported in. *Devon and Exeter Gazette,* 1911.
2. George Gissing (1857-1903), *Born in Exile,* 1892, Part 2, p. 1. Gissing's association with Park Place seems to be coincidental. Pierre Coustillas, the Gissing authority, has pointed out that it is most likely that Gissing would have known of and been sympathetic to William Clifford's philosophical writings. Both Gissing and W.K.C. were involved in the London Sunday Lecture Society programme. Gissing met and became friendly with Lucy Clifford and notes his meetings with her in his diaries.
3. W.K. Clifford, Vol 1, *Lectures and Essays*, Macmillan 1879, Introduction.
4. *Clifford's Genius Shown as a Boy,* Note contributed to the American Mathematical Monthly, 29, 1922, p. 157-158.
5. Autograph Letter Signed=ALS, William Clifford, 1877, Valehouse Collection.

The Cambridge Years
1863-1871

> I want to take up my cross and follow the true Christ, humanity;
> to accept the facts as they are, however bitter or severe,
> to be a student and a lover, but never a lawgiver.
> (William Clifford to Lady Pollock. 1871)

The best impression of William Clifford's early days in Cambridge is that given by Frederick Pollock in the 1879 Introduction to Clifford's *Lectures and Essays,* which he and Leslie Stephen prepared for publication after William's death. Pollock and Clifford both went up to Trinity College in 1863 and became friends when they were third year students. For each of them the friendship became a major influence in their lives. Pollock had first heard of Clifford very early in their first term. He wrote:

> Not many weeks of my first year had passed when it began to be noised around that among the new Minor Scholars there was a young man of extraordinary mathematical powers, and eccentric in appearance, habits and opinions. He was reputed, and at the time with truth, an ardent High Churchman.

Clifford's early High Church religious convictions and belief in Catholic theology were a result of his family background, the teaching he had received at Templeton's Academy, and his studies of St Thomas Aquinas at King's College, London. Pollock reminds us how singular such a stance would have been in those days when the effects of the Church of England's evangelical revival led by Charles Simeon of King's College were still very strong in Cambridge. It was not in Clifford's nature to remain a silent witness to his faith. Pollock describes how William's reputation grew as a result of his lively participation in religious debates through which he tried to find ingenious ways to reconcile advances in scientific thought with traditional Christian beliefs. As the furious discussions raged over the ideas propounded by Darwin in the *Origin of Species,* Clifford found his early formal Christian faith becoming eroded. He was also influenced by the writings of the early proponent of evolutionary theory, Herbert Spencer, and he began his friendships with Thomas Huxley and John Tyndall. His scepticism grew and, by his third year, he had become an eloquent and zealous advocate of agnosticism and, later, of atheism. Edward Carpenter, the social reformer and poet, wrote of meeting Clifford at Cambridge:

> I belonged to one or two little societies which used to meet and discuss literary and other topics. At one of these, which W.K. Clifford organised, I used... to take part in the Sunday evening readings of Mazzini's *Duty of Man*; illustrated by a plentiful accompaniment of claret-cup and smoke! Clifford was a kind of Socratic presiding genius at these meetings – with his Satyr-like face, tender heart, wonderfully suggestive, paradoxical manner of conversation, and blasphemous treatment of existing gods. He invented just at that time a kind of inverted Doxology which ran:
>
> > O Father, Son and Holy Ghost –
> > We wonder which we hate the most.
> > Be Hell, which they prepared before,
> > Their dwelling now and evermore!
>
> and his influence, combined with that of Mazzini, was certainly part of my education at that period.[1]

This blasphemous ditty would have been pounced upon by protagonist and antagonist alike. It says much for Clifford's boyish charm that he could shake off the weight of disapproval that such irreverence must have incurred. As his views changed he spent less time on divinity and the classics, and concentrated more on higher geometry and philosophy. Not that he had neglected mathematics previously, for in 1864, one of his longest papers, *Analytical Metrics*, was written and published in the *Quarterly Journal of Pure and Applied Mathematics*.

He was actively interested in a wide range of subjects. Besides literature, history and the modern languages, he studied Arabic, Greek and Sanskrit, and, because he was interested 'in all methods of conveying thought', he learned shorthand and Morse code. He certainly made an early impression on Frederick Pollock, himself no mean scholar, with his remarkable ability to cut through conventional treatment of mathematical problems. For an example of this ability, Pollock cited Ivory's Theorem, a proposition in the analytical treatment of statics concerning the attraction of an ellipsoid. This was tackled in the textbooks of the day by using a formidable array of co-ordinates and integrals, which achieved the solution without imparting any real understanding of the proposition. He recounts how Clifford described, in one memorable conversation, the geometrical conditions on which the solution to Ivory's Theorem depended and effortlessly made it crystal clear.[2] This solution was not just the provision of a useful geometrical picture; it involved a recognition that the mechanical problem could be expressed and solved in geometrical terms, and prefigured his later belief that the whole of physics could be reduced to geometry. Fred Pollock, who was the same age as Clifford, came to love him 'as his own soul' and, after William's death, kept his wife Lucy as one of his closest friends. In 1868, the year that both he and Pollock were elected Fellows of Trinity, William, together with Fred and his brother Walter and some other friends, visited Dresden during the long vacation. He loved foreign travel, and it was to be a sad consequence of his illness that most of his subsequent travels were undertaken in the search for health and strength. During these early Cambridge years the few surviving letters to his father and step-mother, whom he called Dearest Mama, demonstrate the warmth of his affection for them and his concern for their health and that of his young step-brother and sisters. [3]

Clifford made an impression on all who met him; soon after his arrival in Cambridge he received an invitation to join the Cambridge Conversazione Society, founded in 1826 and known as The Apostles. Frederick Pollock's father had been a member and he recalled this description of the Apostles:

> Their number was very few, and their mode of election was the most remarkable I have ever known. The vacancies were exceedingly rare – perhaps one or two in the course of the year – and the utmost care and study were bestowed on choosing the new members. Sometimes months were given to the consideration of a man's claim. Rank neither told for a man nor against him. The same with riches, the same with learning and, what is more strange, the same with intellectual gifts of all kinds. The same too with goodness; nor even were the qualities that make a man agreeable any sure recommendation of him as a candidate .... The man was not to talk the talk of any clique; he was not to believe much in any of his adventitious advantages, neither was he to disbelieve in them – for instance to affect to be radical because he was a lord. I confess I have no one word which will convey all that I mean; but I tell you that above all things he was to be open-minded. When we voted for a man we generally summed him up by saying, 'He has an Apostolic spirit in him' . . . no honour ever affected me so much as the being elected, as a youth, into that select body . . . and some of the foremost men of the time belonged to that society.[4]

Frederick Pollock goes on to fill out the picture of the Apostles and the way in which his own membership of the Society enriched his life. He mentions Clifford among the glittering array of members as one who, 'if the fates had suffered it, would have been in line with Einstein as his companion or maybe precursor'. The Society, as its members knew it, has a fascination and a mystique about it that has attracted attention over the years. For many of those favoured with election it became the most important influence in their lives. Henry Sidgwick, the great nineteenth-century moral philosopher, described it as a group through which members acquired 'a belief that we could learn, and a determination that we will learn, from people of the most opposite opinions'. William Clifford clearly enjoyed belonging to the group. At the meetings one member would speak to a proposal, and the fundamental rule of honest thought and speech, would guide the open discussion which followed. On leaving Cambridge the status of 'Apostles' changed; they became 'Angels'. After 'taking wings' they could still attend the meetings and the annual dinner. In 1871 when Clifford had rooms in Whewell's Court, he sent notice of the next Apostles meeting to Lord Houghton, who was at that time one of the most prestigious 'Angels'. Clifford jokingly referred to Houghton as the 'Most Apostolic' and announced that he proposed to discuss the subject: '*Is rebellion the whole duty of Man?*'[5]

Besides election to the Apostles, William was winning prizes for his academic work, but he counted it a greater honour when his gymnastic and athletic feats drew praise. In 1869 he wrote:

> At present I am in a very heaven of joy because my 'corkscrew' was encored last night at the assault of arms: it consists in running at a fixed upright pole which you seize with both hands and spin round and round descending in a corkscrew fashion.[6]

When he took his degree it was noted in the university journal, *Bell's Life*, that intellectual distinction and manly exercises were not incompatible since 'the second Wrangler was also one of the most daring athletes in the University'. One can imagine the charm of this brilliant student, able to gain the highest academic honours and yet so boyishly proud of his physical prowess and the acclaim it received. There was more bravado to come; a gymnast friend reported:

> His nerve at great heights was extraordinary. I am appalled now to think that he climbed up and sat on the cross bars of the weathercock on a church tower and when by way of doing something worse I went up and hung by my toes from the bars he did the same thing.[7]

This interest and involvement in physical exercise was not in any way eccentric in undergraduate life in early and mid-Victorian times in Cambridge. There is much recorded fact and anecdotal detail about the athletic prowess and physical feats of eminent mathematicians during their Cambridge years.[8] The heavy academic demands of the Tripos could bring about mental exhaustion and many candidates, including, in 1856, Henry Fawcett, achieved lower placing than expected because of this stress. Academic tutors and coaches encouraged the idea that physical and mental fitness were linked. By the 1840s regular physical exercise was established as an integral part of the would-be senior wrangler's diet. James Clerk Maxwell, who was second wrangler in 1854, regularly carried out extremely vigorous – even violent – feats involving running and swimming. Leslie Stephen, twentieth wrangler in 1854, was renowned at Cambridge for his prowess as a runner and long-distance walker and went on to ambitious climbing expeditions of Alpine peaks which are recorded in his publications about mountaineering in the 1860s. But there was danger in the extreme belief that mind and body could always be strengthened by experiments which tested conventional limits.

William Clifford seemed at that stage to be able to keep the balance between physical and mental activity and to be set for a glittering career in Cambridge when he was brought to a head-on collision with University authorities because of his philosophical views.

At that time all Scholars were required to affirm their allegiance to the Church of England by publicly signing each year the Thirty-nine Articles, which had been drawn up three centuries earlier. William had signed three times, although he had had serious doubts in 1865, and had come to feel that by agreeing to sign he had compromised his integrity. In 1866 he refused to sign. Despite his religious views he was nevertheless elected a Fellow of Trinity in 1868. Others were concerned about academic freedoms too, and, in 1868, thirty-two Fellows of Trinity signed a petition against required religious affirmations in academic life. Eventually, in 1871, Gladstone set up a reform of the statutes, and the break away from the domination of the entire University by the clergy began to take place. Also, while Clifford was a Fellow, he become involved the struggle to bring about reforms to the entire teaching structure of the Tripos. As a result, electromagnetic and thermodynamic topics were included in the curriculum for the first time.

Cambridge students attaining a first-class degree in the Mathematical Tripos examinations are titled 'Wranglers'. In Clifford's day the Wranglers were listed in order, and to be 'First' or 'Senior' Wrangler was a prestigious honour. To reach this high standard it was usual to work with a coach and follow a long and arduous course of specific problem-solving leading up to the examination. But William Clifford had wide-ranging interests. He refused the academic straight-jacket of working only on examination subjects as suggested by The Reverend Percival Frost who was his personal tutor. He chose to study the fresh, original works of J.J. Sylvester, Arthur Cayley and G. Salmon and their great continental contemporaries. In Pollock's words he 'omitted most of the things he ought to have read, and read everything he ought not to have read'. Yet, in spite of almost no preparation, his powers of original thought won him, like J.J. Sylvester, William Whewell, Clerk Maxwell and Sir William Thomson in their time, the place of Second Wrangler. He also shared first place in the other glittering award for mathematicians, the Smith's Prize. Clifford later wrote to Fred's mother, extolling the advantages of holding second place – 'much more secure than the First. It takes a King indeed to despise the Vizier!'[9]

* * *

While at Cambridge Clifford was a member of another prestigious group, the Grote Club. Other members were the famous economist Alfred Marshall, Henry Sidgwick the philosopher and John Venn the logician. Even among these giants, Clifford sparkled, and years afterwards Marshall would remember the brilliance of his discussions but remarked that 'he was too fond of astonishing people'.[10] This fluency and power had brought Clifford to prominence in his third year when he won the Trinity College declamation prize. He chose as his subject Sir Walter Raleigh and delivered his speech in the form of a dramatic dialogue; the original manuscript still exists among his personal papers. Clifford led his audience through the life of Raleigh from the age of twenty-four, and he used the story to enunciate his beliefs.

> Now, as then, there is a Dorado, meant for the good of all men, the gift of Him who sends rain upon the just and upon the unjust. The student of science lives in the consciousness that at any moment that may be revealed to him which shall change utterly the whole face of society, and alleviate in an enormous degree the physical miseries of mankind.

And now, as then, there is the danger lest that which is meant for the good of all should be perverted into an instrument of evil; lest, after all, the only result should be that another portion of conquered Nature is cursed for the sake of man.[11]

This success resulted in his being asked to deliver the annual oration at the Commemoration of Benefactors. His brief then was to honour the recently deceased Master of Trinity, the famous Dr Whewell. Clifford again managed to make an original and unexpected approach. His opening words for the eulogy were 'Thought is powerless except it make something outside of itself: the thought which conquers the world is not contemplative but active. And it is this that I am asking you to worship today.' He used this allegory to support his thesis:

Once upon a time – much longer than six thousand years ago – the Trilobites were the only people that had eyes; and they were only just beginning to have them, and some even of the Trilobites had as yet no signs of coming sight. So that the utmost they could know was that they were living in darkness, and that perhaps there was such a thing as light. But at last one of them got so far advanced that when he happened to come to the top of the water in the daytime he saw the sun. So he went down and told the others that in general the world was light, but there was one great light which caused it all. Then they killed him for disturbing the commonwealth; but they considered it impious to doubt that in general the world was light, and there was one great light that caused it all. And they had great disputes about the manner in which they had come to know this. Afterwards another of them got so far advanced that when he happened to come to the top of the water he saw the stars. So he went down and told the others that in general the world was dark, but that nevertheless there were a great number of little lights in it. Then they killed him for maintaining false doctrines: but from that time there was a division amongst them, and all the Trilobites were split into two parties, some maintaining one thing and some another, until such time as so many of them had learned to see that there could be no doubt about the matter.[12]

The whole performance made a great impression on the academics present, and William developed and extended this theory of the intellectual growth of mankind in his essay *On Some of the Conditions of Mental Development,* first given as a lecture at the Royal Institution in 1868. Clifford's tribute to Dr Whewell was remembered by two young women who had attended the declamation at Trinity. They were nieces of Samuel Page Widnall, a well-known and much loved resident of Grantchester, a picturesque village near to Cambridge. Clifford had become something of a favourite with this family and in the late 1860s and early 1870s he was a frequent visitor at their beautiful home The Old Vicarage. He is remembered in family diaries as a most lively young man who was fond of entertaining the children with tricks and puzzles. Alice and Amy Smith liked to visit Cambridge – chaperoned as was the custom in those days – and at different times in 1866 they attended a Union debate, a Trinity

William Clifford in 1868, photographed by the keen amateur photographer, Samuel Page Widnall, probably in the garden of The Old Vicarage, Grantchester, outside Cambridge.

College chapel service, and a display in the Gymnasium in which Clifford may well have been a participant. One of their friends, who became a professional artist, Charlotte Mary Greene, an aunt of Graham Greene, remembered William Clifford's irreverent sense of humour. One day in King's Parade, as she was admiring a picture by Turner in a shop window, Clifford approached her and said: 'I can tell you how that picture was painted. The artist squeezed a lot of his paints on to a canvas and sat on it!' In old age, Polly Greene as she was known, wrote a nostalgic poem in the style of Rupert Brooke's *The Old Vicarage, Grantchester,* and again remembered Clifford with these lines:

> Who sits and writes beneath the pines?
> 'Tis Clifford making nonsense rhymes,
> For Euclid's science here would seem
> Unfitted for this home of dream.

She also remembered his nonsense contribution to an amusing pamphlet sold to gather funds at the church bazaar. It ran

> Among the things not generally known is this – the report of a pistol, seen at the distance of five ounces, smells like the taste of half an hour, only it is not white.

Samuel Page Widnall's practical, artistic and literary talents included photography, and his striking image of Clifford, probably taken in 1868, shows Clifford with no signs of the illness that would prematurely end his life.[13]

William did not lose his delight in frivolous verse and the following example, titled 'On coming in sight of the sea from the hills above Porlock', was composed a few years after he left Cambridge, and was found amongst his personal papers.

> This, this, I said with deep emotion
> Is what is called the briny Ocean.
> Propelled alike by tail and fin
> The little fishes swim therein.
> The great ones also find their quarters
> In this expansive waste of waters.
> Its bosom, which the wild winds whip,
> Is sailed upon by many a ship,
> And steamers also you may view
> Which go by paddle wheel or screw.
> In fact there's nothing half so grand
> Excepting what is called the land
> And that receives a different name
> Because it is not quite the same.
>   Signed W.K. Clifford.

While at Cambridge, William met and became a close friend of Henry Fawcett, a follower of John Stuart Mill, who, though blind, had in 1863 been appointed Professor of Political Economy. They were both active members of the Republican Club and their friendship continued after William moved to London.

1870 was an exciting year for Clifford. He was very much looking forward to going to Göttingen where he was to meet the mathematician R.F.A. Clebsch, but that trip never materialised. However, in the summer vacation he and his college friend George Crotch, with whom he had enjoyed various gymnastic and athletic escapades in Cambridge, set off round Europe for some weeks. William wrote to Fred's mother about their adventures. They travelled informally and walked a lot of the way through France and Spain. They climbed the Pic du Midi

Royal Astronomical Society commemorative photograph of the 1870 Eclipse Expedition with William Clifford seated at the far right.

to view the range of the Pyrenees having walked 'over twenty miles, with rise of 10,000 feet'. They thoroughly enjoyed their casual wanderings, living on five francs a day, getting drunk on cheap wine, observing and sketching butterflies and buildings, and sometimes sleeping rough.[14]

The previous December, quite suddenly, the government had put up the money to finance a Royal Astronomical Society Expedition to Sicily to view the 1870 eclipse of the sun. William was invited to be one of the team. He wrote from Florence to Lady Pollock, after the ship, the Psyche, carrying the scientists and their equipment struck rocks and was wrecked off Catania. The spelling has not been corrected.

> No ink, no paper, no nothing – well if ever a shipwreck was nicely and comfortably managed without any fuss – but I can't speak calmly about it because I am so angry at the idiots who failed to save the dear ship. At Catania, orange groves and telescopes; thence to camp at Augusta, great fun, natives kept off camp by white cord; 200 always to see us wash in the morning – a performance which never lost its charm – only five seconds totality free from cloud, found polarization on moon's disk . . . . At Rome two and a half days, pictures, statues, Coliseum by moonlight . . . . This morning arrive in Florence – Pitti Palace – spent all my money . . . . Addio.[15]

The team had to struggle, penniless and hungry, back to Ostend. Clifford writes: 'I saw the corona; its breadth was about an eighth or tenth part of the moon's diameter, on the side where the sun was about to emerge. I also caught a glimpse of a prominence but the time was too short to pay any attention to it.' The Society's report concluded that 'photographs taken at the time with a rapid rectilinear photographic lens, showed great extensions and were considered specially successful'. In fact, they were fortunate to have been able to make any observations at all since there were comparatively dense, low-lying cumuli in the region.[16]

During his seven years at Cambridge Clifford had dramatically evolved from precociously devout Anglo-Catholic to radical critic and fierce enemy of all organised religions. Although some of the constricting religious requirements at the University had been relaxed, Cambridge could not be a comfortable place for one who would later pronounce that 'There is only one thing in the world more wicked than the desire to command and that is the will to obey.' Fortunately, University College, London, which had been founded in 1828 and become fully incorporated into the University of London in 1836, had, unlike its sister college King's, no religious restrictions in its statutes. Jews and Dissenters who had previously been precluded from graduating were welcomed at what was familiarly known as the 'Godless Institution of Gower Street'. For this freedom and for other academic reasons William Clifford now turned his eyes to London.

Notes

1. Edward Carpenter, *My Days and Dreams,* George Allen and Unwin, 1921.
2. F. Pollock, Introduction, Vol 1, *Lectures and Essays*, W.K. Clifford, Macmillan, 1879.
3. W.K.C's papers in the Valehouse Collection.
4. Sir W.F. Pollock, quoting from Sir Arthur Helps in *For My Grandson*, John Murray 1933, p. 30.
5. ALS, Valehouse Collection.
6. As note 2.
7. ibid.
8. Andrew Warwick, *Exercising the Student Body*. Chapter eight, 'Science Incarnate', C. Lawrence and S. Shapin eds, University of Chicago Press, 1998.
9. ALS, February 1870, University College London Archive, Folder MS add. 136.
10. John Maynard Keynes, *Essays in Biography*, (On Alfred Marshall), London 1933.
11. As note 3.
12. W.K. Clifford, *Lectures and Essays*, Vol 1, Macmillan 1879, Introduction.
13. Photograph of W.K.C. and the Polly Greene anecdote contributed by Lady Jennings of Grantchester.
14. As note 2.
15. ibid.
16. The Royal Astronomical Society *Memoirs*, Vol. 41.

'The Great Scientific Missionary':
University College, London
1871-1875

> No simplicity of mind, no obscurity of station, can escape the
> universal duty of questioning all that we believe.
> (William Kingdon Clifford: *Ethics of Belief*)

While he was at Cambridge, Clifford's brilliant academic lectures, including his discourses at the Royal Institution in 1868 and 1870, had made him well known in London. In 1869, he was elected to the committee of the London Mathematical Society, which had been established in 1865. There he would have met the eminent mathematical physicist, James Clerk Maxwell, who had been elected in 1867. Earlier still, when he was a young student at King's College London, Maxwell was professor in the Department of Applied Sciences there from 1860 to 1865 and Clifford would have attended his lectures on Natural Philosophy and Astronomy. Thomas Hirst, who was Professor of Mathematical Physics at University College, was keen to attract William to the vacant Goldsmid Chair of Applied Mathematics there and Clerk Maxwell supported William's application. He wrote a glowing testimonial:

> The peculiarity of Mr Clifford's researches, which in my opinion points him out as the right man for a chair of mathematical sciences, is that they tend not to the elaboration of abstruse theorems by ingenious calculations, but to the elucidation of scientific ideas by the concentration on them of clear and steady thought. The pupils of such a teacher may not only obtain clearer views of the subjects taught, but are encouraged to cultivate in themselves that power of thought which is so liable to be neglected amidst the appliances of education.[1]

After 1874 Clifford and Maxwell often met as Fellows of the Royal Society. They shared a mutual belief in the importance of Hamilton's 'quaternion' methods, and used them in their teachings and writings. Later, religious differences would complicate their relationship and their attitudes to fundamental research.

To further attract William to London, the University Council persuaded Sir Francis Goldsmid to add an extra £200 a year to his original 1868 endowment, and William took up his new post as Professor of Applied Mathematics and Mechanics in the Spring of 1871.

Professor Edwin Power in his article *Exeter's Mathematician*,[2] records that Clifford was a most successful professor, an excellent teacher, loved by his students, and that 'his syllabuses in applied mathematics could well stand today in the classical branches of the subject.' Clifford arrived at University College to find that men and women were taught in separate classes and, in order to limit intermingling of the sexes, men's classes changed on the hour and the women's classes on the half-hour. Women were even banned from crossing the front quadrangle and, even in London, it was not until 1878 that women gained the right to obtain degrees. William was the first of the mathematicians to admit both sexes to his lectures. Many educationalists at that time were agreeing that women might safely be allowed to follow courses of instruction, but believed that encouraging them to challenge conventional thinking could only lead to disruptive discontent. However, Clifford wrote:

> It seems to me that the thing most wanting in the education of women is not the acquaintance with any facts, but accurate and scientific habits of thought, and the courage to think that true which appears unlikely. And for supplying this there is a special advantage in geometry, namely that it does not require study of a physically laborious kind, but rather that rapid intuition which women certainly possess.[3]

It is interesting to note that, Ellen Watson, the first woman to enter the mathematics classes at University College was considered by Clifford to have a 'level of proficiency very rare in a man.'– he noted that he was 'totally unprepared to find it in a woman.' Indeed the notes that she took of some of the lectures proved to be invaluable when Clifford's *Mathematical Papers* were being prepared for publication after his death.[4]

Clifford was quickly surrounded by a large circle of friends. Carey Foster, Alexander Williamson, Edward Frankland, George Croom Robertson, Joseph Hooker and G.J. Romanes were academic colleagues in London. His links with Thomas Hirst, Thomas Huxley, John Tyndall, Leslie Stephen, Arthur Cayley, J.J. Sylvester and Henry Fawcett had been established during his time at Cambridge, and through these connections he was now plunged into the very centre of intellectual life in London. Frederick Pollock had already started his legal career in London. His father, Sir William Frederick Pollock and his wife welcomed William into the colourful circle of personalities from the theatrical and artistic world, as well as academics, politicians and lawyers who were their friends. Lady Juliet Pollock, whom, in his letters, William had addressed as 'Chère Mère Noble' and signed himself as 'thy child', was devoted to William. She and her two sons, together with Moncure Conway and his wife, were very frequent visitors to the evening meetings in his bachelor rooms in St John's Wood. William wrote some of the loveliest of his letters to her, and one senses that she took the place of the mother he lost so early in his life.[5]

Moncure Conway was an American from Virginia who met William Clifford at Cambridge and was captivated by his personality and influenced by his free-thinking. He wrote of his friendship with William and applied to him Goethe's definition of genius as 'having an immense capacity for work' rather than as possessing mysterious powers. Conway had turned his back on his Methodist pro-slavery family upbringing and had become a Freethinking abolitionist. He moved to London and set up the South Place Institute for Advanced Religious Thought. He sought William's company and wrote of him: 'He was gay of temperament, fond of merry society, indefatigable in the dance, accomplished in all manner of pretty tricks and surprises.' It was Conway who described William Clifford as the 'great scientific missionary'.[6]

When he was just twenty-two William Clifford had used these words in his first lecture at the Royal Institution in 1868:

> If the mind is artistic, it must not sit down in hopeless awe of the monuments of the great masters, as if heights so lofty could have no heaven beyond them. Still less must it tremble before the conventionalism of one age, when its mission may be to form the whole life of the age succeeding . . . . To become crystallised, fixed in opinion and mode of thought, is to lose the great characteristic of life, by which it is distinguished from inanimate nature: the power of adapting itself to circumstances.[7]

This visionary style of address from the young mathematician-philosopher became characteristic of Clifford, and he developed into a most powerful speaker. At the Brighton meeting of the British Association in 1872 when he spoke on *The Aims and Instruments of Scientific Thought,* his discourse was astonishingly eloquent and Conway writes of a 'pure and beautiful light' diffusing from his face as he spoke these closing words:

> When a poet finds he has to move a strange new world, which his predecessors have not moved, when, nevertheless, he catches fire from their flashes, arms from their armoury, sustentation from their foot-prints, the procedure by which he applies old experience to new circumstances is nothing greater or less than scientific thought. When the moralist, studying the conditions of society and the ideas of right and wrong which have come down to us from the time when war was the normal condition of man, and success in war the only chance of survival, evolves from them the conditions and ideals which must accompany a time of peace, when the comradeship of equals is the condition of national success; the process by which he does this is scientific thought and nothing else. Remember then that it is the guide of action; that the truth which it arrives at is not that which we can ideally contemplate without error, but that which we may act upon without fear; and you cannot fail to see that scientific thought is not an accompaniment or condition of human progress, but human progress itself.

William's mercurial personality and winning eloquence coupled with his high intellect made him welcome wherever intellectuals gathered. He did not limit his involvement simply to the academic world. He was elected one of the youngest Fellows of the Royal Society,[8] but he volunteered to present a series of ten lectures on Geometry to women in South Kensington. He was a Fellow of the London Mathematical Society, but he would speak on Sundays at St George's Hall where the Sunday Lecture Society's programme would cover a broad range of philosophical subjects for a wide audience.

In December of 1874 he gave four lectures at the Town Hall in Shoreditch for a University extension course. As was his habit, Clifford did not make detailed notes of the lectures, but a shorthand reporter recorded them at the time of delivery. These lectures were published by Macmillan in 1879 in the volume: *Seeing and Thinking*. The diagrams were provided by Professor Michael Foster FRS whose *Primer of Philosophy* was recommended reading for those following the course of lectures. The four sections are titled: *The Eye and the Brain, The Eye and Seeing, The Brain and Thinking* and *Of Boundaries in General*. The lectures are full of practical examples to illustrate the physiological, anatomical and physical subject-matter. Macmillan were publishers of inexpensive books in the *Nature Series,* and *Seeing and Thinking* was a companion volume to others by eminent scientists of the day. There was Sir John Lubbock on *Flowers, Fruits and Leaves,* Sir George Gabriel Stokes on *Light,* and *Popular Lectures and Addresses* by Sir William Thomson. Clifford's lectures are easy to read and must have been even easier to follow verbatim when we remember his gift for captivating his audiences. His closing words at the final lecture contain this warning:

> I dare say, now, that you are rather indignant at being kept so long hearing perfectly obvious remarks that are true of everything. You may think that it is beneath the dignity of human nature to spend all this time in contemplating the size and shape of a piece of wood. Very well; it is written in the fifteenth chapter of the Koran that when Adam was created all the angels were commanded to worship him. But Eblis, the chief of them, refused, saying, 'Far be it from me that am a pure spirit, to worship a creature of clay.' And for this refusal he was shut out forever from Paradise. Now the doom of Eblis awaits you if you fail to give due reverence to these little obvious everyday things – things that are true of every stone that lies on the pavement, of every drop of rain that falls from heaven, of every breath of air that fans you. Like him, you will find with astonishment, that the creature of clay that you despise is the Lord of Nature and the Measure of all things, for in every speck of dust that falls lie hid the laws of the universe; and there is not an hour that passes in which you do not hold the Infinite in your hand.

Seeing and Thinking was a popular book and was reprinted four times.

Clifford's aim was to express the most complicated mathematical and philosophical ideas in simple language, and his talks were well attended. For example, at his three lectures on *Philosophy of Pure Science*, published in *Lectures and Essays* and first given in the afternoons at the Royal Institution in March 1873, the audiences numbered 165, 212, and 143 – high numbers for that type of lecture. Of course it was an exciting age with Darwinian theories provoking open questioning of established ideas. A book on the history of the

Metaphysical Society, which was formed in 1869, is sub-titled *Victorian Minds in Crisis,* and that title aptly describes the scenario into which William Clifford's lectures were launched.

Within two years of his arrival in London he had delivered several of his best-known lectures and published, amongst other things, his famous paper on biquaternions, which was concerned with the motion of bodies in curved space. However, only one piece of Clifford's work, *Elements of Dynamic, An Introduction to the Study of Motion and Rest in Solid and Fluid Bodies* was prepared in book form during his lifetime.

It was after his death that his friends and colleagues, gathering together whatever of his notes and papers they could find, edited and produced six further books. The task was not an easy one. One of the problems was that Clifford was a spontaneous and inspired speaker who preferred to work with only the very minimum of notes, sometimes even writing his notes after the lecture. His memory was phenomenal, but some of the notes were never completed. The fact that he avoided accepting the seriousness of his illness meant that he did not have time to make sure that his papers were in order before his death. H. J. S. Smith and R. Tucker took up the task of preparing the mathematical works for publication, Fred Pollock and Leslie Stephen jointly did the philosophical ones, and Karl Pearson edited and completed his popular work on science.

The first of these posthumous publications, in 1879, was the two-volume *Lectures and Essays*. Clifford had intended to recast this collection of lectures of a non-specialist mathematical nature into a book to be called *The Creed of Science,* but it fell upon the shoulders of the editors to gather and present them after his death. They explain and account for any changes or exclusions that they have made, but for the most part they are exactly as William Clifford wrote them. The book also contains full reference to when and where the lectures were first delivered. Four of the lectures were originally given to the Sunday Lecture Society. William would have delighted in this Society's catholic aim – 'to deliver lectures on Science – physical, intellectual and moral, History, Literature and Art, especially in their bearing upon the improvement and social well-being of mankind'. The lectures were given in St George's Hall, Langham Place and were enormously popular. Six other lectures were first heard at the Royal Institution. This prestigious institute had been set up by an American, Count Rumford, in 1799. The Institution, which occupies grand premises in Albermarle Street in London, has no students, but four Research Professors are appointed and part of their duty is to give lectures on the results of their work to the membership and, at times, to the public. The evening discourses have become famous and somewhat fashionable occasions when certain rituals of address are followed. They have attracted a mainly academic and, at least in Victorian and Edwardian times, wealthy audience. In those days evening dress would be worn by the principal guests and the lectures were a glittering social, as well as academic, event. John Tyndall, President of the Royal Institution in Clifford's day, was a very close friend to William and later to Lucy. She used one of the evening discourses as a venue in her 1910 novel, *Sir George's Objection.* The first of the series of Christmas Lectures for Children, which are now a television favourite, were given by Michael Faraday at the Royal Institution in 1826.

Macmillan published *Lectures and Essays* as 'by the late William Kingdon Clifford, FRS, late Professor of Applied Mathematics and Mechanics in University College London and sometime Fellow of Trinity College, Cambridge'. Pollock's magnificent Introduction with its Biographical, Bibliographical and Letters sections, provides a full and most personal picture of William Kingdon Clifford, and Pollock regretted only, on agreeing to write the biography, that it was going to be impossible to do full justice to his friend. He wrote:

> The discourses and writings collected in this book will indeed testify to the intellectual grasp and acuteness that went to the making of them. Clifford's earnestness and simplicity, too, are fairly enough presented to the reader, and the clearness of his expression is such that any comment by way of mere explanation would be impertinent. But of the winning felicity of his manner, the varied and flexible play of his thought, the boundless range of his human interests and sympathies, his writing tells – at least, so it seems to those who really knew him – nothing or very little.

Pollock goes on to state that he had taken on 'the higher responsibility of telling the world that it has lost a man of genius; a responsibility which must be accepted even with the knowledge that it cannot be adequately discharged'.

Then come the eighteen lectures and essays. A glance at their titles gives an idea of the formidable range of subject material, and on the frontispiece, the Paul-Louis Courier maxim: *'La vérité est toute pour tous'*, is witness to Clifford's guiding principle. Among the contents of Volume One are: *On Some of the Conditions of Mental Development; Atoms;* and his review articles: *The First and Last Catastrophe* and *The Unseen Universe*. The lecture entitled *The Philosophy of the Pure Sciences*, has been singled out as most helpful to 'readers who are not mathematicians by profession but who ... from a general sympathy with all branches of intellectual activity are disposed to follow new mathematical ideas'. In it Clifford analysed, 'in popular phraseology but with the utmost scientific precision' the fundamental principles of geometry and arithmetic. The especially important essay *On The Aims and Instruments of Scientific Thought*, first delivered to the members of the British Association on 19 August, 1872, contains one of the many instances in which Clifford anticipates Einstein's concept of the curvature of physical space. William's public lectures, often *'pièces d'occasion'* delivered at academic meetings, thus formed the basis of most of his published work.

Clifford was active socially, too. Sir Sidney Colvin, art critic and Director of the Fitzwilliam Museum, who became a lifelong friend of the Cliffords, writes in his memoirs of an interesting club. Originally called 'The New Club' and later renamed 'The Savile', it had been set up in 1868 to gather together people who wanted to break away from the ritual snobbishness and formality of most London clubs. William was elected to the club in 1869 while he was still at Cambridge. Among the founding members were John Morley, Lord Houghton, Richard Holt Hutton and Leslie Stephen. They set up the Savile as a club where members expected to chat freely amongst each other without traditional formal introduction. Colvin had met Clifford at Cambridge and described him as:

that short-lived genius unequalled and unapproached in the rarefied region of speculation where higher mathematics and metaphysics merge into one . . . with beautiful lucidity of mind and mastery of style . . . extremely striking and attractive, with his powerful head and blunt Socratic features, the candid almost childlike upcast look of his light grey-blue eyes between their dark lashes.[9]

Clifford had become friendly with Robert Louis Stevenson and, in 1874, he supported Sidney Colvin in proposing him as a member of the Savile. They were naturals for the spirit of such a group and by their sparkling personalities and lively conversation contributed much to the success of the Club. In 1874 Clifford wrote:

At the Savile I found C., who had just done dinner, but sat down while I ate mine, and we solved the universe with great delight until A. came in and wanted to take him off to explain coins to somebody. Of course I would not let him go . . . . We walked about in the New Road solving more universe!

He went on to outline their discussion about the need to seek earnestly after truth and to accept only provisionally opinions one has not inquired into – a theme to which he returned again and again. There was much discussion about Mazzini's idealistic and revolutionary views on the improvement of social organisations, and Clifford spoke with an almost missionary zeal of the need for 'some earnest person to go and preach around in a simple way the main straightforward rules that society has unconsciously worked out . . . how truth and right are to be got at by free enquiry and the love of our comrades for their own sakes and nobody else's'.[10]

In addition, Clifford was president of the Dialectical Society and also a member of the Republican Club. He was overseeing the scientific papers for *The Academy, A Fortnightly Record of Literature, Learning, Science and Art*; the cross-cultural element of this publication would have particularly appealed to him. As well as serving on the Council of the London Mathematical Society he was secretary, and then vice-chairman, of the Mathematical and Physical section of the British Association. But of all his London involvements, it was his membership of the Metaphysical Society that gave William Clifford the greatest pleasure. The Society had been founded in 1869 by Alfred Tennyson, James Knowles, editor of the *Contemporary Review* and founder of the *Nineteenth Century*, and Charles Pritchard, Savilian Professor of Astronomy at Oxford. They met each month and, as with the Apostles, one member was invited to present a paper for discussion. In fact, three of Clifford's papers to the Metaphysical Society were later published in his *Lectures and Essays*. This society, which was active for eleven years, had originated because of the intellectual conflict thrown up by Darwinian theories. Its broad aim was to attempt to reconcile by discussion religious belief and the findings of science. There was a balanced membership between Senior churchmen, atheists, Roman Catholics, theists and freethinkers.

Only Jews and Mohammedans were not represented. Among the members were Cardinal Manning, Thomas Huxley, Walter Bagehot, F.D. Maurice, John Ruskin, John Tyndall, William Ewart Gladstone, Arthur Balfour and Henry Sidgwick. The only rule of discussion was that the views of the other members must be respected, and this rule must have been tested to its limits, considering how deeply seated were the differences of opinion held by the members.

* * *

In 1873 William met and fell in love with Sophia Lucy Jane Lane. She was described as a golden-haired, red-cheeked art student who sketched antique statues in the British Museum.[11] At that time she was beginning to make her mark as a writer and had had three or four stories serialised anonymously in *The Quiver*. William and Lucy became engaged to be married in 1874 and in that year she wrote a twenty-page letter to William voicing some of her doubts and fears about his atheism. She could never before have met anyone even faintly like him: so powerfully committed to seeking the truth, so intellectual, so uncompromising, so loving. She confessed her belief that the divine origin of Christ might be true by asking:

> Won't it be a good thing if some good is done in his name that would not be done for the sake of ordinary men, out of sympathy and comradeship? And is it not natural and likely that he should have set apart certain men to preach this same doctrine, and have given them some of the same wonderful power?[12]

William was gentle with her and told her that, although he had ceased to believe in the supernatural goodness of Jesus, she, his darling, if she had such a belief, 'should keep it, since a person about whom we know so little is perhaps the safest sort of figure to clothe your ideal'. However, he did not spare her his views about the priesthood. His seven closely written pages take her through his notion that no Christian ecclesiastical body has ever had the power to persecute without using it. When it had been pointed out to him that, while the Quakers in Pennsylvania had the power to persecute their Indian servants, they didn't use it, he rejoined that the Quakers had no official clergy. He spoke of past times and of the clergy sapping the foundations of social life with monasticism and the theological motive, getting the hospitals suppressed and the physicians banished, and substituting places 'where a martyr's toe in a silver box was brought to cure you', shutting up philosophical and scientific schools and 'reducing all Europe to a black night of barbarism which Greece had not known for two thousand years'. He wrote:

> Even now the clergy hinder education of children except by their own formula, knowing well that a straight conscience and a free-grown intellect will neither believe in their doctrines, nor approve their precepts . . . . The priesthood has destroyed one civilisation and has just failed to strangle another at birth and that it is up to every honest man to see that it shall never have another chance.[13]

Poor Lucy! These were strong words – especially in the year 1875. One of her brothers was a parson, her sister was in a convent and she was in love with one of the most outspoken atheists of the age.

Perhaps William did not actually intend this for the eyes of his 'sweet child', for he ends: 'There's a sermon for you! Poor little thing, there is one comfort, that you won't read it. Farewell my own child. I shall see you at eleven tomorrow.' Whether he delivered the letter himself, or talked her through it later, it does seem likely that she read it anyway, for it was found in William Clifford's papers when he died. He was uncompromising in his atheism, but Lucy's consolations from her love for William were many, and not least of them his translation from Heine which he sent to her:

> Sweet, when I look at you
> Into your eyes
> All my pain, fearing you
> Passes and flies.
> Love, when these lips of mine
> Burn upon yours,
> Ah, what a health divine
> Into me pours.
> When I lean silently,
> Dear, on your breast,
> Then there comes over me
> Heaven's own rest.
> When your soul tells me you
> Love me, at last;
> Darling, my tears will flow
> Bitter and fast!

William wrote Lucy beautiful letters with entertaining accounts of his doings. In these two verses of a philosophical poem, he attempts to express his beliefs:

> I hold you this once more to my heart
> And the sweet swift seconds roll away;
> Tomorrow – how many miles apart!
> – How near today?
> The summer dies out, sun by sun;
> The lily droops to the ground and dies;
> Dies; but the root in the ground lives on
> That shall one day rise.
> Is it thus with me, O sun of my days?
> Shall death lay hold on me, after you,
> Till you shine again, and the fresh warm rays
> Revive me too?
> The old tales tell of a soul of things,

> How earth and sky are made of his breath,
> How in one man's flesh he folded his wings
> And died the death.
> All my world is of one love made;
> Earth and sky are the limbs thereof;
> Life and death are its light and shade,
> And the soul is Love.
> When I leave my darling, he veils his grace,
> But lives incarnate in earth and sky;
> Then dies into glory before her face,
> And reigns for aye.[14]

While visiting friends in Cambridge this great man, who could captivate any audience at any level, wrote to Lucy that he was missing her, 'because I cannot shine and fascinate anybody without you to help. That is because you are such a darling.' This mutual dependence and respect grew stronger during the eighteen months of their engagement.

During the years 1867 to 1878 Clifford was producing and delivering the mathematical works that were later to be formally published by Macmillan as *Mathematical Papers*. One of his dying wishes was that Robert Tucker, who was secretary to the London Mathematical Society, should see the manuscript through the press. On completing the massive task Tucker's moving letter to Lucy, who had encouraged and assisted him, contained the words 'I can now only wish that the result of my efforts may deserve your approval, and may not be unworthy of the illustrious mathematician, your much beloved husband.'[15] The book came with a fold-out sheet of mathematical notes in Clifford's own hand. Tucker also writes a preface, which includes a delightful anecdote about Clifford as a little boy. It recounts how he dealt with a complicated puzzle: – a solid wooden snowball-sized sphere which, with great difficulty and a great number of movements, could be taken to pieces. William did not touch it. He simply looked at it, put his head in his hands for about ten minutes and then, to everyone's amazement, disassembled it without hesitation. The mind of a geometric genius had been at work. Following this and other stories of Clifford's youthful exploits comes the thirty-six page *Introduction* to the papers written by H.J.S. Smith, Savilian Professor of Geometry in the University of Oxford. Professor Smith includes in his introduction the words:

> Clifford was above all and before all a Geometer . . . and geometry was to him an important factor in the problem of 'solving the universe' . . . . It is true that in the treatment of geometrical questions he shows a marked preference for symbolical methods; and, as might be expected, a marvellous command over analytical expression . . . . But, whatever the method employed . . . the properties of space remain the perpetual subject of his contemplation.

Professor Smith ends:

How much may have perished unrecorded we cannot tell. But, however this may be, no geometer will look for a more splendid monument of Clifford's genius, or for a more touching memorial of his early death, than is to be found in the unfinished pages: On the Classification of Loci which embody the last and perhaps the greatest effort of his inventive powers.

Then come the fifty-one papers themselves. These form the mathematical meat of Clifford's work and remain as the principal monument to his exceptional intellect. Most of these papers deal with various branches of geometry; some are concerned with curved spaces, which Clifford regarded as physically important. Several papers developed what are now known as Clifford Algebras; they are: *Preliminary Sketch of Biquaternions,* 1873; *Further note on Biquaternions,* 1876; *On the Classification of Geometric Algebras,* 1876; and *Applications of Grassmann's Extensive Algebra,* 1878. Dirac's theory of the electron, was based on the Clifford Algebra appropriate to four dimensional space-time which is fundamental to modern physics and which is more fully described in the chapter The Clifford Heritage. Dirac invented this algebra in the 1920s, but it is very closely related to Clifford's algebra of biquaternions.

Clifford was, of course, interested in other people's work, and he often reviewed contemporary publications. One of his reviews caused reverberations at the time of writing and repercussions after his death. In 1875, Peter Tait, Professor of Natural Philosophy at Edinburgh, who produced among other things a mathematical theory of the dynamics of golf, wrote with Balfour Stewart: *The Unseen Universe or Speculations on a Future State.* In it, they distorted William Thomson's ingenious idea that atoms were vortex rings, in order to give a description of life after death, and thus reconcile physics and religion. Of course, Clifford seized upon this. He wrote a review, later included in *Lectures and Essays,* which criticised the authors' logic and cruelly parodied it.[16] Clifford wrote

> our authors 'assume, as obviously self-evident, the existence of a Deity who is the Creator of all things'. They must both have had enough to do with examinations to be aware that 'it is evident' means 'I do not know how to prove'.

Clifford's essay also contained some of his most important conceptions, including the unity of molecules and the fields in the space between them. The review, with its biting agnosticism and sarcasm, was not destined to win friends. Clifford was unable to write on religious questions without using language that was offensive to some people. His friend John Morley, editor of *The Fortnightly Review,* who later became Viscount Morley and Secretary of State for Ireland, suggested to him that his excessive language weakened his attack. Morley wrote that it would be better to 'imitate old Isaac who put the worm on the hook tenderly, as tho' he loved it'.[17] Tait was stung by this trenchant attack and later caused Lucy distress when he published unfavourable remarks about Clifford's *Elements of Dynamic.*

Beside the various lectures to the Metaphysical Society and other addresses and discourses which Clifford gave to enthusiastic audiences, he contributed reviews and articles to journals such as *Nature, Mind, Contemporary Review* and *Nineteenth Century Review*. One academically significant item, which William Clifford translated from German, was published in *Nature* in 1873. This was the seminal Riemann paper, *On the Hypotheses which lie at the Bases of Geometry*, first published in 1868, which gave a general formulation of the idea of 'curved space'. It led to William Clifford's ideas in 1870 and ultimately to Einstein's General Theory of Relativity in 1915-17. Clifford was the first to present Riemann's work in England and it made a tremendous impact. Another of his most important addresses was to the Cambridge Philosophical Society as early as 1870. He presented a paper which critics find difficult to believe had been produced forty years before Einstein had announced his theory of gravitation. J.J. Sylvester had previewed this work as 'some remarkable speculations' at the 1879 Exeter meeting of the British Association. It is titled, *On the Space Theory of Matter* and is included in *Mathematical Papers*. In it, Clifford presented the idea that matter and field energy are simply manifestations of curvature of space. This concept is discussed more fully in Part Three of this book. It has led many mathematicians to state that in this paper Clifford actually anticipated Einstein's general theory of relativity. One hundred years later, the ideas he expressed in this paper became the focal point of a Clifford Centennial Meeting at Princeton.

In seeking to present the impact that Clifford's written words had at the time of their appearance, there is an impressive array of contemporary reviews to consult. These few words from the review of *The Common Sense of the Exact Sciences* in *The Athenaeum*, of 11 July, 1885, perhaps encapsulate the quality that was Clifford's trademark:

> Clifford was admitted on all hands to be the most remarkable mathematician of his generation.... There is a marvellous charm about Clifford's writing. He has a singular faculty of presenting difficult truths in words of startling clearness and brevity.

This, together with the judgement published in *Nature* after his death that Clifford was 'one of the deepest thinkers and most brilliant writers this century has seen', and mentioning his 'ready sympathy, his delicate sense of humour, and sweetness of disposition', give us perhaps the closest chance of appreciating William Clifford's compelling combination of academic genius, personal charm, and gift for inspirational teaching.

Notes

1. W.K. Clifford, *Lectures and Essays*, Vol. 1, Macmillan 1879, Introduction.
2. Professor Edwin Power, 'Exeter's Mathematician', *Advancement of Science*, (London) 26, 1970.
3. W.K. Clifford, *Lectures and Essays*, Vol. 1, *On Some Conditions of Mental Development*.

4. W.K. Clifford, *Mathematical Papers*, edited by R. Tucker, Macmillan 1882. Preface, p. xxvii.
5. Extracts of some of Clifford's letters to Lady Pollock are included in the Introduction to *Lectures and Essays*, others are in the Valehouse Collection.
6. Moncure Conway, *Autobiography, Memories and Experiences*, Cassell 1904.
7. William Clifford, *On Some of the Conditions of Mental Development,* later published in *Lectures and Essays,* Vol. 1, p. 104.
8. Clifford was elected Fellow of the Royal Society in June 1874 when he was just 29 years old.
9. Sir Sidney Colvin, *Memories and Notes of Persons and Places*, Arnold and Co. 1921, pp. 119-120.
10. ALS, Valehouse Collection.
11. Leon Edel, *The Life of Henry James*, Rupert Hart-Davis 1972, Vol. 5, p.106.
12. ALS, Valehouse Collection.
13. ibid.
14. ibid.
15. W.K. Clifford, *Mathematical Papers.* Edited by Robert Tucker. Macmillan, 1882.
16. P. Tait and Balfour Stewart: *The Unseen Universe or Speculations on a Future State,* Macmillan 1875. W.K.C., *The Unseen Universe, Lectures and Essays* Vol. 1, p. 228.
17. ALS, Valehouse Collection, dated 26 May, 1875.

Life in London:
Friendship with George Eliot
and George Lewes

On 7 April, 1875, Lucy and William were married at St George's register office, Hanover Square. After the honeymoon in Cambridge, Lucy found herself pregnant and their daughter, Ethel Lucy, arrived on 26 January, 1876. The couple had set up home at 26 Colville Road near Notting Hill Station in Bayswater, and they regularly invited friends and academic colleagues there on Sunday afternoons. William's Cambridge friend, Frederick Pollock, was already married. He, his wife, and his parents were most welcoming and supporting to Lucy. Thomas Chenery, editor of *The Times*, Robert Louis Stevenson, Sidney and Frances Colvin, the John Colliers and the Moncure Conways were all regular visitors at the Cliffords' home. Edward Clodd, the rationalist Freethinker, took great pleasure in these gatherings at Colville Road where he first met Thomas Huxley. At the Clifford's he also met Mark Pattison, the Grant Allens, Thomas Hardy, Mrs Lynn Linton, Cotter Morrison, and York Powell. He describes those Sunday afternoons as being linked with 'fragrant and refreshing memories':

> At the Clifford's you were sure to meet someone worth the knowing. There was no Smart Set to fill their empty time and waste yours in inane gossip; no prigs to irritate you with their affectation; no pedants to bore you with their academic vagueness, but just a company of sane and healthy men and women, gentle and simple, who wanted to meet one another and have a full and free talk.[1]

As well as lecturing to his students and carrying out administrative duties at University College, Clifford was preparing the mathematical and philosophical lectures that were later to form the basis of his published works. Between mathematical research, academic commitments, family and social involvements, he somehow found time to write a fairy tale. *The Giant's Shoes* appeared in *The Little People,* a collection of tales by Lady Pollock and Walter Herries Pollock published in 1874. It was judged to be 'a feeble collection of fairy tales redeemed only by a nonsensical story by Professor Clifford, which is wildly fantastic, with no sort of narrative line, but which is very funny.' One of the characters was 'a rare hand at making guava dumplings out of three cats and a shoe horn which is an accomplishment rarely met with'.[2]

Also on the lighter side, he and Moncure Conway joined together on escapades aimed at exposing the forms of superstitious fraud which were taking place at Spiritualist meetings. The pair of them would go to séances with the intention of disrupting the proceedings. A famous medium of the time, Mrs Guppy, fell victim to their pranks. With Clifford and Conway seated on either side of her she was inhibited, and eventually routed, by William's insistence that a set of false teeth from his overcoat pocket should materialise on the table. Another famous medium, a certain Mr Williams, came to Conway's house to hold a séance. However, Clifford and Conway took seats beside him in the darkened room and refused to release his fingers and allow him to perform his tricks. He fled the house in confusion leaving some of his apparatus behind him. He left the country and had his hoaxing paraphernalia seized by the Customs officials.[3] Lucy wrote that William loved to entertain their friends with hilarious dramatisations of these events.

It was in 1875 that Clifford dictated two full chapters and parts of several more with the intention of producing a book to be titled *The First Principles of the Mathematical Sciences Explained to the Non-Mathematical,* but he did not finish the work. Shortly before his death he chose a new title, *Common Sense of the Exact Sciences,* and left instructions about how he wanted it to be completed. Eventually, in 1885, Macmillan published it. It has become a classic, and the five bold chapter titles, *Number, Space, Quantity, Position,* and *Motion* indicate the range of attack. At first, the production of this volume was in the hands of William's colleague at University College, Mr R.C. Rowe. Unfortunately Rowe died in 1884, and Karl Pearson, then aged twenty-seven, and who was himself to become a most eminent mathematician, was asked to complete the task. It was not easy. Several chapters were in proof form and the corresponding manuscript pages had been destroyed. There were confusions about the placing of some sections of the chapters, and the early plan made by Clifford was only discovered after the book had been completed. As well as editing, Pearson had to make considerable changes, and he himself wrote half of Chapter III and the whole of Chapter V. He agreed to do this because he thought that, even 'completed by another hand we can only hope that it will perform some, if but a small part, of the service which it would undoubtedly have fulfilled had the master lived to put it forth'. In tackling this editorial task, Karl Pearson, who took on the task with 'a grave sense of responsibility', was thrown into close collaboration with Lucy Clifford. Their joint desire to do what William had intended formed the beginning of a friendship that lasted all her life. There were contemporary new editions of some of his books, but *The Common Sense of the Exact Sciences* and *Mathematical Papers* are the only of William Kingdon Clifford's works to have been republished in recent times.

Mathematical Papers came, textually unaltered, from Chelsea Publishing Company, New York in 1968. Earlier, in 1946, Alfred A. Knopf of New York published a new edition of the *Common Sense of the Exact Sciences,* with an introduction by James R. Newman and a preface by Bertrand Russell. Both men have interesting comments to make about William Clifford. Bertrand Russell first read the book with 'intoxicated delight' when he was just fifteen. He states that:

> Clifford possessed an art of clarity such as belongs only to very few great men . . . clarity that comes of profound and orderly understanding, by virtue of which principles become luminous and deductions look easy.

He surveys the mathematical ideas of the book and goes on to speak of the importance and relevance of Clifford's philosophical thinking – even fifty-seven years after his first reading of the book. He writes:

> Clifford was much more than a mathematician: he was a philosopher of considerable merit in what concerned the foundations of mathematical knowledge . . . he saw all knowledge, even the most abstract, as part of the general life of mankind, and as concerned in the endeavour to make human existence less petty, less superstitious, and less miserable.

He goes on to give this simple example which is relevant to all teaching:

> A wood in which trees are planted in rows looks regular when viewed along a row from one end of it, but may appear higgledy-piggledy when viewed on a slant. The same sort of thing is true of a mathematical subject. If you approach it from the wrong angle, each step will be difficult, you will be entangled in thickets, and you will get no view of the whole; but if you start at the right point and advance in the right direction, the obstacles disappear and progress is easy. Clifford's survey of elementary Euclidean geometry beginning with the two axioms that things can be moved without change of shape, and that the size of things can be increased or diminished by a change of scale without change of shape, is just what is needed to make geometry easy to a beginner.

Russell's *Preface* is another affirmation of Clifford's gift of clarity, and of his dedication to explaining modern scientific thought to the uninitiated in the simplest way possible. The book was published in London a year later and reprinted in New York in 1955. James R. Newman's forty-eight page *Introduction* to this edition contains much from Frederick Pollock's earlier biographical account and also includes more recent appraisals of the impact of Clifford's work.

In their years together in London the Cliffords were at the centre of a group of the most exciting and influential people of those times, but their friendship with George Eliot and George Henry Lewes was particularly precious to them. In 1845, the year that William Clifford was born, George Henry Lewes had published his *Biographical History of Philosophy*. Lewes had extremely wide-ranging academic interests. He published books on literature, drama, philosophy and popular science, and, in addition to translations from German and countless contributions to journals, he had founded the *Fortnightly Review* in 1865 and was its editor for two years. His marriage was not a happy one and he lived apart from his wife and children. From 1854 to the end of his life he and George Eliot lived together as man and wife. The young William Clifford was among the distinguished group of friends surrounding this famous couple. Lucy recalled that:

> They knew my husband before I did. They used to say that he was more like a happy schoolboy than a mathematician already famous in the scientific world; they delighted in his unexpectedness, his daring

flights of thought and the soft voice in which, accompanying himself, he sometimes sang to them a setting of some verses by Thackeray that I never came across except among a small group of Cambridge men.[4]

William Clifford wrote a review for *Academy* of G.H. Lewes's series: *Problems of Life and Mind* which was published between the years 1873 and 1879. This review delighted Lewes, who declared that it had: 'gratified me more than anything else written about my book before. Pages of laudation would be trivial compared with such appreciation.'[5] The two men shared the same philosophical outlook and Matilde Blind, the feminist writer, who knew them both, noted that they also shared the gift of being able to make complex problems intelligible. They also both delighted in jokes and funny stories. George Lewes reported on an occasion when Lucy and William had been at the house for tea. William had told an amusing story about the Bible being translated into Zulu. A difficulty had arisen in finding a suitably equivalent word for God. At last a word was chosen, but one of the natives present explained that the word also meant 'meat in a state of decomposition.' This was mentioned at the Savile Club, when a visitor, raising his glass to his eye, said 'Ah! No doubt a translation of "The Most High!" '[6] At another time William displayed his sharpness of tongue in his description of an acquaintance whom he felt was an unscientific philosopher: 'He is writing a book on metaphysics, and is really cut out for it; the cleverness with which he thinks he understands things and his total inability to express what little he knows will make his fortune as a philosopher.'[7]

People were naturally curious, and generally censorious, about this remarkable woman writer who was openly living with a family man who was not her husband. Lucy herself had first-hand evidence of the effect that public opprobrium had upon George Eliot. When he was single, William had been a regular visitor to the Lewes' home. He knew that George Eliot approved of his marriage to Lucy and yet, when he visited after his marriage, no mention was made of his bringing his new wife with him. He came to realise that George Eliot would not risk a snub, and so never invited anyone who had not intimated beforehand that they would like to come to the house. As soon as he mentioned that Lucy would be honoured to meet the Lewes's, she was invited. Lucy's memories of being taken to *The Priory* were poignant, and she recalled them in the article which was described by Marie Belloc Lowndes as the 'most living presentment of that singular woman and of George Lewes during the last years of their union.' Lucy takes the reader alongside her into the hallowed residence, and shares the domestic details that, perhaps, other observers would not include:

> Consider what a time it was: the great poets and men of science of the century were alive – there were many spaces in Westminster Abbey that have since been filled: the Metaphysical Society was in full swing, the Nineteenth Century was in the making. Luckily there was George Henry Lewes. The grip of his hand, the cordial, almost merry greeting, carried many a quaking stranger through the first awkward moments . . . . He was considered to be a very ugly man: why, I cannot think,

for his expression was so pleasant, so kindly, that it distinguished his features – if they were bad . . . . He made one think of a dog in many ways – a rather small, active, very intelligent dog. I think he must have worn slippers . . . for one never heard his footsteps; and he pattered about after the manner of a dog: he was so easily pleased, too, and showed it in a way that made one insensibly wonder if a nice fringy tail were not wagging behind him . . . . If he talked with you apart . . . he usually spoke of George Eliot as *She*; you felt that the first letter was a big one . . . . Sometimes he came to our house . . . and it was a shock one day when, in relating some anecdote that amused them, he alluded to George Eliot as Polly. All my life I had thought of her as a wonder, and had spoken her name after a little hush, as he seemed to do when he used 'She' to indicate her. And to hear her called Polly! . . . Does anyone who went there live and for get? To go even to one of the Sunday afternoons was felt to be an honour . . . . Interviewers and strangers of every description were kept at bay. On the left (in the small double-drawing room) as you faced the fire, with her back to the window, sat George Eliot . . . . On her right, and more or less in a circle round her sat the visitors. They seated themselves as noiselessly as possible after a reverential greeting to their hostess. People didn't break up into groups and talk with each other in those days . . . . But each one spoke in turn and was listened to by the others . . . it was a terrible ordeal for the average intelligence. If you happened to be frightened out of your wits (as I was) you felt a choking sensation in your throat. Then perhaps you managed to get something out in a squeaky voice. I merely said that 'I thought so too' or something to that effect. But Mrs Lewes probably helped you through with a kindly expression on her wonderful face. Wonderful? Yes, and like a horse's. Her likeness to Savonarola has often been mentioned: you were sensible of it the moment you saw her first; but you were also immediately reminded of a horse, a strange variety of horse that was full of knowledge, and beauty of thought, and mysteries of which the ordinary human being had no conception . . . . And her fascination, her magnetism, the exquisite thrill that went through you at the sound of her low measured voice, at the sight of her little generally undeveloped smile, like a fitful gleam of pale sunshine, was beyond all description, and had the effect of making you feel that there was nothing you would not do for her . . . . At five o'clock tea was brought in . . . . Mr Lewes poured out the tea . . . . Mrs Lewes was reverently served, and we all sat solemnly down with cups in our hands and bits of doubled thin bread and butter in our saucers. Then talk was resumed . . . literature, philosophy – a great deal of philosophy – painting, and occasionally, not often, music . . . . The talk was a little sententious, a little too good, so that it seldom had an air of spontaneity . . . the best things were generally said by George Eliot herself. In an argument that greatly interested her, now and then her beautiful hands contributed something to the scene that it lacked, something natural, a suggestion even of poetry, or of passion; not often for it was an astonishingly well balanced, dignified group. On one occasion we stayed behind for a little while. My husband had been ill; we were going abroad in a few days, and not likely to see her soon again. When we four were alone – the Leweses and ourselves – she took my hand and held it tightly while she talked to my husband; and, when we were going, she kissed my cheek and said 'God bless you, my dear' with something in her voice that made my heart bound and tears come to my eyes. Perhaps she divined, although we did not, that it was the last time we should ever go there – or see them together. A few days later she and George Henry Lewes came to see us, but unhappily we were out: we never quite got over it – and they never came again: it was never possible . . . . George Henry Lewes died in November 1878 and my husband early in 1879.[8]

Beside his academic work and his connection with his direct colleagues, Clifford was forming a wide range of friends in other fields. On a trip to Paris with Lucy in 1876, they met up with his friend Eliza Lynn Linton. She was a highly respected journalist, novelist, essayist and social critic. She admired William immensely and wrote to Lucy when she heard that he was seriously ill: 'It is not saying more than I feel when I say that willingly – willingly, I would give my life for his.'[9] These were strong words; but William inspired such powerful sentiments and he was always recognised as an exceptional, even prophet-like, character. He regarded the formalism of religions as subversive to morality and one of his favourite sayings was 'A way with ghosts and phantoms, the Kingdom of Man is at hand.' In the notes he had prepared for the Congress of Liberal Thinkers he wrote:

> As to the priestly organisation, 'the Church' has always been adverse to morality, and is now . . . . And I believe that, so far as the Christian organisation is concerned, the time has come for heeding again the ancient warning: 'Come out of her, my people, that ye be not partakers of her sins, and that ye receive not of her plagues.'[10]

These views demonstrate what Richard Holt Hutton, the powerfully academic editor of *The Spectator*, called Clifford's 'almost measureless irreverence'. William often quoted from Swinburne and Tennyson, and was a great admirer of Walt Whitman's *Leaves of Grass*. He enjoyed composing frivolous ditties but he was a serious poet too, and wrote these undated lines for Lucy:

> I went out desolate in the night
> And hoped no help of the solemn skies,
> But the stars shone kindly, and in their light
> I saw your eyes.
> I plunged all reckless into the sea;
> The waves came threatening, crest on crest;

> But they heaved so lovingly under me,
> I knew your breast.
> I held all men for an alien race,
> Strange to me, seeking another goal;
> My brothers helped me, and in that grace
> I loved your soul.[11]

William would return home to Lucy elated after meetings of the Metaphysical Society and she wrote that he would be:

> full of glee if the Bishops had appeared and he had been in good form. He would not only tell me everything everybody had said but mimic the manner in which it was said. We have sat over the fire and shouted with laughter when he added ridiculous little tags of his own to what had really been said.[12]

Lucy goes on to tell how, after one meeting, William arrived at the station and found he had no money. Lord Arthur Russell turned up and William borrowed a shilling from him, bought a ticket, and then paid back 6d on account! After his death, knowing that Fred Pollock was engaged in writing some biographical notes, Lucy wrote to him asking that William's 'spice of wickedness' should be portrayed in case people thought that he lived always 'at high pressure like a prophet'.

Many of the evenings at the Savile Club and meetings of the Metaphysical Society and other groups would have been in smoke-filled rooms, and the way to and from them, at least in winter, would have been through fog-filled streets. Hardly surprisingly, the respiratory problems that had begun to develop in William's early days in London worsened. His friends were aware of this, but, as we see from his reply to Huxley's warning that he should take a rest from his academic work in July 1876, Clifford did not realise how ill he was:

> I think that you are the truest friend in all the world. . . . But I'm sure that to give up my college work next term would do me more harm than good . . . . I should continually fret about one or two things I want to finish, and rest from work is no good at all without rest from worry.

He reminded Huxley that, later in the year, he and Lucy would have seven weeks sailing in the Mediterranean and then, after time in the Pyrenees, he would be:

> as strong as a horse before the next cold weather. I don't believe that too great or too sudden a change of life can be good except in very bad cases, and there is really not much the matter with me. Besides that, it is enough to make any man well at once to think that he has such friends.[13]

He was totally and tragically wrong.

The 1876 trip to the Mediterranean had been set up and financed by friends. Huxley and Pollock, realising that without a period in a warmer climate, Clifford was in mortal danger, had, without his knowledge, written to some of his friends and colleagues and asked for financial help. The readiness with which the contributions came in indicates the esteem and affection Clifford inspired. The renowned mathematician J.J. Sylvester whom William knew well through the London Mathematical Society, sent a contribution to the *'Conspiracy Fund'* as he called it. Like many of the other contributors he offered more money should it be needed. He had taken the Tripos in 1837, when the University was still a Church establishment and, although he was brilliant enough to be ranked second Wrangler, was nevertheless disqualified from being awarded a degree because he was a Jew and could not therefore subscribe to the Articles of Faith. One of Sylvester's main mathematical interests over several decades was in polynomial invariants and covariants. In 1878 he published a paper based on a diagrammatic representation of these mathematical entities, making a bold and highly imaginative analogy between these diagrams and the structure of chemical molecules. Independently, Clifford also developed a systematic graphical representation of invariants and covariants. This was eventually published, in 1881, as *Mathematical Fragments*. This is a remarkable and rare volume. Only fifty copies of the foolscap-sized volume were produced. It consists of facsimiles of Clifford's unfinished papers misleadingly entitled *Theory of Graphs*. The delightful thing is that it is reproduced entirely in Clifford's own hand, and shows exactly how he worked. After Clifford's death Sylvester visited Lucy and sent her copies of some of his verses. As far as is known he did not send her his legendary 400 lines rhyming with Rosalind!

James Clerk Maxwell sent his contribution of £5 – offering more if needed – along with a witty three-page letter that included these words:

> I hope that a slight displacement of his position on the earth's surface may bring him into a milder air and one less stimulating than that of Gower Street, so that, as his oscillations between elliptic and hyperbolic space gradually subside, he may find himself settling back again into that parabolic space wherein so many great and good men have been content to dwell, and may long enjoy the three treasures of the said good and great men as enumerated by S.T.C. [Samuel Taylor Coleridge 'Three treasures, – love and light, and calm thoughts, regular as infants' breath . . . . '][14]

In June of that year Lucy left her baby daughter in London and travelled with William. She was to use her memories of these travels to advantage in her later books. But, during these early days of her marriage, Lucy kept working at her writing. Each edition of *The Quiver* from 1871 to 1877 carried a story of hers in instalments. For these early writings Lucy drew on the backgrounds that she was familiar with. The stories are set in suburbs like St John's Wood and Shooter's Hill where she and William had been based before marriage. They mainly follow the

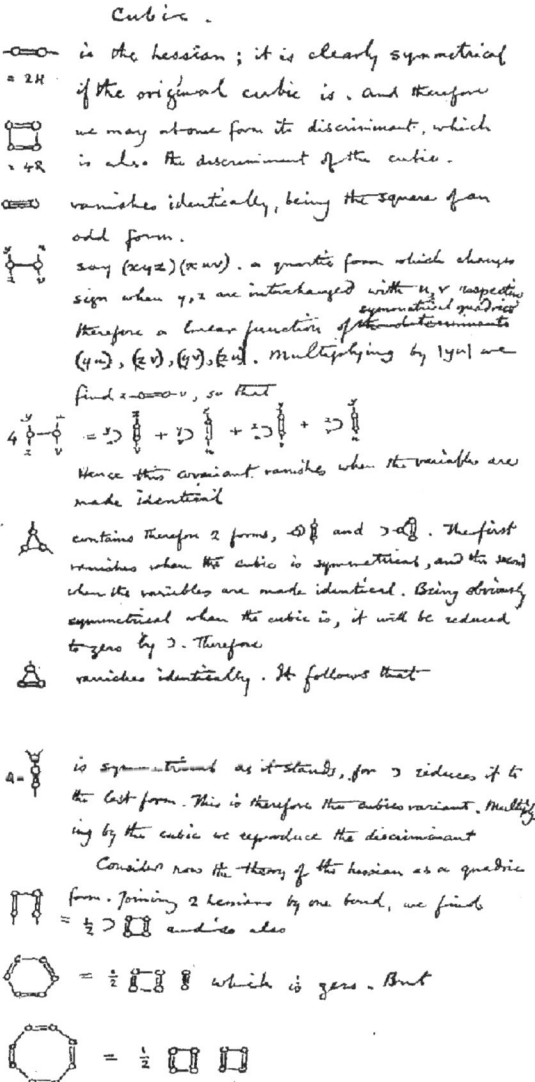

Page 2 of Mathematical Fragments. *This rare facsimile edition of William Clifford's papers was published in 1881. When he came across this book about ten years ago Sir Roger Penrose discovered that Clifford used a diagrammatic notation in the theory of invariants very similar to one he thought he had invented himself.*

same formula and deal with middle-class or newly poor families, often with girls having to make their livings as governesses, teachers and artists. There are lost letters, dying fathers, misunderstandings between lovers and mistaken identities, all of which get happily sorted out in the concluding pages. As with most of the fiction in *Quiver* at that time there were moralistic overtones with the characters ending up wiser and better for the events and confusions of the tale. Lucy later developed some of these serialised stories as novels. The Cliffords had a modest home and could afford to employ only a cook and a live-in housemaid. Much of the burden of housekeeping, entertaining friends and looking after William would have fallen on Lucy's shoulders and, with the arrival of Alice Margaret their second daughter on 11 April, 1877, Lucy would have had a struggle to find time for her writing. Yet, in that year, she delivered a twenty-two instalment story to the editors of *Quiver*.[15] She developed, in these early days of her marriage, the disciplined habits of work that were to sustain her and her daughters after William's death.

William was working tremendously hard during 1877 and 1878. Although, during the second year of his marriage, it had been increasingly clear that William was a sick man, the young married couple with their company of friends may in some way have generated enough passion for life that the spectre of early death was pushed to the shadows. Perhaps even George Lewes's earlier words 'that nothing should rob him of his glorious destiny' may have, in some mystical way, seemed to grant him the right to disregard his physical limitations. William set aside the conventional rules of thought and belief and threw himself into the centre of religious and scientific controversy. Frederick Pollock wrote that Clifford's undoing lay in the fact that he had:

> a perilous excess in his own frame of nervous energy over constitutional strength and endurance. He could not be induced, or only with the utmost difficulty, to pay even moderate attention to the cautions and observances which are commonly and aptly described as 'taking care of one's self'.[16]

His very liveliness would prove to be the main ingredient of his physical decline, for he habitually overtaxed his strength. He believed that by occasionally pushing himself to the limit and working all night he was strengthening his resilience. But, in Pollock's words, 'he thought he was making investments when in fact he was living on his capital'. To complete his eighteen-page review essay *The Unseen Universe,* Clifford took up his pen at nine forty-five one evening and worked straight through till nine the next morning.[17] Part of the trouble must have been Clifford's mismanagement of his own time. He was interested and involved in so many activities that he tended to leave the actual writing up of articles till the very last moment. We have an amusing but imperative note to him from John Morley, editor of the *Fortnightly Review*:

> Your complete manuscript must be in our hands by 8am Monday morning .... If all contributors took their ease as you do, what would become of

me, I wonder? When you have five minutes to spare, decide the question whether fidelity and punctuality in small matters are not part of altruistic morals. It is of no use to threaten you with hell-fire![18]

Many of William Clifford's lectures were delivered without full notes; he also had the reputation of being a reluctant letter writer. He must have caused problems at University College, for he hated bureaucracy and especially detested writing testimonials. He once wrote to Fred Pollock:

Seventeen people have written to ask me for prescriptions, I mean testimonials. They know that writing matrimonials drives me mad, that every testimony takes me a week to do, that it sears my conscience and sores my brain – why are people such fiends? They only do it to annoy, because they know it teases. Therefore pity me and persuade others to do likewise.[19]

Unanswered letters were stuffed into his pockets. However, when he did settle down to writing, he produced splendid letters full of vivid and often irreverent descriptions. He had an earthy sense of humour. He wrote from Marseilles that, as a result of too much travel and foreign food, he had 'become as one who hath no excrement, and whose inward parts refuse to make dung'. With the aid of castor oil, he had 'got the better of mine enemies' and had made good use of the *lieux d'aisance* in the city streets – 'mighty convenient and no extra charge for paper'. This was included in a letter to Fred Pollock in July 1876, together with lively descriptions of the countryside, the local peasants and their real Provençal language, and laced with impious comments about the 'confounded church which towers above everything, with a gilt statue on the top'. They climbed up so that 'Notre Dame de la Garde was well below us and could be looked down upon with just contempt.'[20]

In another of his 1876 letters sent from Algeria to Fred Pollock he breaks off the travelogue to tell him that he has sent, to Macmillan the publishers, his ideas for a set of school books. He favoured the idea of short ten-minute lessons following the Pestalozzian plan to make children find things out for themselves. His rather grandiose plans included teaching in different languages on successive days with no direct formal teaching of language until 'there are enough facts to make Grimm's Law intelligible'. He does admit that this method would require a French and a German teacher, both trained and competent, besides the English one! The next day he bought for himself 'twelve volumes of the Bibliothèque Nationale for 3 francs – Rabelais, five volumes, and Montesquieu, Pascal, Diderot and Vauvenargues.' Thinking of these 25-centimes *livres de poche,* he gets swept along with his ideas for a publication of two or three hundred such volumes for all literature, in English. 'Fancy, the Pensées of Pascal with the notes of Voltaire, Fontenelle and Condorcet, a good life at the beginning and all well printed by the million.' some of these letters are included in the introduction to *Lectures and Essays* and are essential reading for all who would know William Kingdon Clifford. Another letter from the 1876 travels is to Lady Pollock. It is full of colourful vignettes. We read of the amiable ship's captain, the amorous Frenchman, the gazelle that tried to eat Lucy's hat, the incredible dancing man, the lectures on Arabic that William attended – so many amusing incidents, and all of them vividly described. From Algeria they went to Malaga, and William's letter to Fred from there is outrageously critical of the Spanish of whom he says that the only one member of that race who *may* possibly have told him the truth, had lost so many teeth that he left out all the consonants and was, therefore, impossible to understand! He and Lucy had day to day problems with hotel owners and taxi drivers and hated having 'to think that every man you meet is ready to be your enemy out of pure cussedness'. In Granada too they were pestered by beggars and threatened by aggressive dogs, but their seven weeks of travel enabled William to regain some strength.

When they returned to London, William again took up his academic pursuits. However, there was a now a new sense of urgency about the way he worked, and his writings – especially those stating his ethical and philosophical views – became much more vigorously outspoken and critical. Earlier in 1875 Clifford's essays *On the Scientific Basis of Morals* and *Right and Wrong* were strongly worded and radical but not violently controversial. But now he produced two new essays *Ethics of Belief* and *Ethics of Religion* both of which were fiercely and destructively critical of contemporary religious thought and practice. The first of these essays caused serious repercussions. It contained the famous conclusion that it is 'wrong always, everywhere, and for anyone, to believe anything on insufficient evidence'. The editor of *Contemporary Review,* Sir James Knowles, accepted and published the essay early in 1877. Alexander Strachan, the owner of the *Review*, strongly disagreed with his action, believing that Clifford's views were too controversial and extreme. As a result Knowles resigned and set up the more impartial and very successful journal *Nineteenth Century*.

The essay, *Ethics of Belief*, in which Clifford placed, without compromise, the responsibility for what we believe on each and every one of us, was published in Volume Two of *Lectures and Essays.* Above all else, the collection demonstrates the clearness and courage with which Clifford dealt with the ethics of intellect. In his essay Clifford introduces the topic of 'The Duty of Inquiry' with a parable:

> A shipowner was about to send to sea an emigrant-ship. He knew that she was old, and not over-well built at first; and that she had seen many seas and climes, and often needed repairs. Doubts had been suggested to him that possibly she was not seaworthy. These doubts preyed upon his mind, and made him unhappy; he thought that perhaps he ought to have her thoroughly overhauled and refitted, even though this would put him to great expense. Before the ship sailed, however, he succeeded in overcoming these melancholy reflections. He said to himself that she had gone safely through so many voyages and weathered so many storms that it was idle to suppose that she would not come home safely from this trip also. He would put his trust in Providence, which could hardly fail to protect all these unhappy families that were leaving their fatherland to seek for better times elsewhere. He would dismiss from his mind all

ungenerous suspicions about the honesty of builders and contractors. In such ways he acquired a sincere and comfortable conviction that his vessel was thoroughly safe and seaworthy; he watched her departure with a light heart, and benevolent wishes for the success of the exiles in their strange new home that was to be; and he got his insurance money when she went down in mid-ocean and told no tales. What shall we say of him? Surely this, that he was verily guilty of the death of those men. It is admitted that he did sincerely believe in the soundness of his ship; but the sincerity of his conviction can in no way help him, because he had no right to believe on such evidence that was before him.

Clifford goes on to the state that we do wrong to ourselves and to Man by allowing ourselves to be credulous when there is the possibility for us to question further and seek truth. He cites as examples the followers of Buddha and of Mohammed: 'both cannot be infallibly inspired; one or the other must have been the victim of a delusion, and thought he knew that which he really did not know. Who shall dare to say which? And how can we justify believing that the other was not also deluded?'

Volume Two contains ten other sections including, *On the Nature of Things in Themselves*; *The Influence upon Morality of a Decline in Religious Belief;* and *Cosmic Emotion*. The most attractive element of the lectures and essays is the gentle way in which the audience, or reader, is led through the sometimes complicated ideas that are presented. There are amusing anecdotes, classical references, practical demonstrations and down-to-earth examples. All of them are introduced in simple language and mirror the popular philosophical interests of the age.

We have another contemporary view of Clifford. In 1877 an anonymous satire on English society and ideas, *The New Republic,* was published in serial form in the magazine *Belgravia*. The book, which was eventually revealed as the work of William Hurrel Mallock (1849-1923) who had won the Newdigate Prize while at Balliol, 'burst like a bombshell' on the intellectuals of the time because it presented easily recognisable portraits of contemporary figures. Mallock formed them, under fictional names, into a group gathered together for a weekend visit to a country house. This device allowed them to group and re-group for discussions about God, the Church, culture and the social élite. Through these discussions and arguments, the prejudices and shortcomings of each of the illustrious visitors are revealed. The three friends Huxley, Tyndall and Clifford are characterised as the scientists Storks, Stockton and Saunders. Saunders is militantly atheistic, and Saunders' hatred of orthodox religion reflects the tone of Clifford's polemic essay *The Unseen Universe.* In one of Saunders' contributions to the discussions we hear him state: 'Science will drain the marshy grounds of the human mind so that the deadly malaria of Christianity, which has already destroyed two civilisations, shall never be fatal to a third.' *The New Republic* was not kind to Clifford. He came over as brash and shrill, and the gentler, comical and humorous side of his personality is disregarded. William, and certainly Lucy, must have been hurt by it. One very outspoken critic of Mallock's book was George Eliot. She wrote:

I think it one of the most condemnable books of the day; not simply because the Master of Balliol is a friend for whom I have a high regard. If I had known nothing of Mr Jowett personally, I should have equally felt disapprobation of a work in which a young man who has no solid contribution of his own to make, sets about attempting to turn into ridicule the men who are most prominent in a serious attempt to make such contribution . . . . I think that kind of direct personal portraiture (or caricature) is a bastard kind of satire . . . .²

Others represented in the satire were Ruskin, Pater, Matthew Arnold and Carlyle. Only Ruskin, as Mr Herbert, comes out well in the end. Mrs Mark Pattison, who through her second marriage became Emilia Dilke and thus the mother-in-law of the Cliffords' elder daughter, was thought to be represented in the book as Lady Grace. However, that direct representation is disputed by some critics. When *The New Republic* came out William Clifford had been reading George Eliot's *Daniel Deronda* and he wrote about it to Lucy:

One feels that one is looking at things with a large-minded sympathetic companion who is great enough to take in the best side of all the people she describes. It is exactly the opposite to that poor creature Mallock, who catches superficial traits of men one knows to be great, and makes them mean.[22]

1878 saw William Clifford finishing the single paper defining Clifford Algebra. This seminal work had been first presented as, *On the Classification of Geometric Algebras*, an abstract for the London Mathematical Society in 1876. But it was extended and developed and sent across the Atlantic to J.J. Sylvester under the title *Applications of Grassmann's Extensive Algebra*. Sylvester was at that time professor at Johns Hopkins University. He published Clifford's paper in the new *American Journal of Mathematics* of which he was the founding editor. The concepts in the papers are derived as a generalisation and union of the ideas of Hamilton and Grassmann, who, in the early 1840s in apparently different contexts, had introduced a new kind of multiplication. Grassmann had, by 1877 (the year in which he died), arrived at the Clifford algebra structure, though in a different and less simple way. It is intriguing to note that Grassmann may well not have known of Clifford's 1876 abstract or Clifford of the Grassmann 1877 paper.

1878 was Voltaire's centenary year. To mark the occasion Moncure Conway, William Clifford and Thomas Huxley had planned to set up a *Congress of Liberal Thinkers* to propagate liberal ideas and opinions. Moncure Conway saw William Clifford as a 'reappearance of Voltaire' and certainly Clifford, with his positive hatred of superstition, was set to *écraser l'infâme*. A powerful committee was formed and Huxley was elected President. On 13 and 14 June, nearly four hundred delegates attended the first congress in London. It was held at South Place in Islington – the meeting place established and run by Conway. There were important representatives, including women, from Europe and from America. Broad Churchmen, Unitarians, secularists, theists, atheists and Hindus, were among the delegates. In addition there was vigorous support from intellectuals of the time who were unable to attend. The event was a resounding success, with the venerable George Jacob Holyoake, who had invented the term 'secularism' and had been imprisoned for professing his atheism, addressing the enthusiastic group who were meeting together for the very first time. The only cloud over the proceedings was that William Clifford's health had failed and he was unable to attend. This decline had been accelerated by the news of the death of his father in February 1878. William was too ill to attend the funeral and yet he engaged on one last furious effort to produce what was eventually published as the very last item in *Lectures and Essays*. In April, while under doctor's orders to take absolutely no risks, he again worked right through the night to produce one of his best and most readable treatises, *Virchow on the Teaching of Science*. Pollock saw this as a tremendously impressive academic achievement but a possibly fatal exertion. The strategies discussed in it concerning the differences between 'being told' and 'finding out' are still relevant today. William even attempted to get back to his academic teaching, but he collapsed while trying to finish a lecture. No human constitution could operate at that pitch without serious repercussion. Huxley had once light-heartedly commented to Lucy 'You let that old man be out at night too much – I caught him at a late hour with a Bishop – he gets into bad society!' But now, recognising that William could kill himself with this sort of overwork, it was Huxley who intervened. He formally sought the advice of doctors, friends, and family and then booked passages for William and Lucy to the Mediterranean. He wrote to tell Lucy their decision:

> Absolute rest is to be his portion. A set of irons will be supplied to you in which he is to be put, if physically refractory. Also a bottle of chloroform wherewith any attempt at serious thinking, or indeed the making of a joke . . . are to be immediately stopped . . . the one thing that Willie is to avoid is fatigue – I cannot impress this too strongly upon you. There must be no landing and toiling about to see interesting things in the sun – no exposure at night – a day's imprudence may ruin a month's building up.[23]

Later that month, leaving their tiny daughters to be cared for by friends in London, William and Lucy again set off to spend some months in warmer climates. On board the *Morocco* from Fiume to Malta, William managed to scribble some notes for the opening address to the *Congress of Liberal Thinkers*, which he vainly hoped he would return to attend. Their travels, organised and paid for by the fund set up by Leslie Stephen and Thomas Huxley, took them to Malta, Genoa, Milan, Lake Como, and then to a sanatorium at Monte Generosa in Switzerland where many consumptives had sought cures. William sent amusing drawings of himself being carried up the mountain on a type of sedan chair mounted on the shoulders of four bearers with Lucy ahead of him perched on a mule. They hit a period of quite exceptionally harsh weather there, and Lucy wrote to the Pollocks complaining that they might freeze to death in the 'coldest, most shivery, chatter-your-teeth, sort of place you could possibly imagine'.

The Pollocks offered to bring Ethel and Margaret out to Switzerland to be with their parents but Lucy and William decided to return to London. Now, however, William's health was broken beyond repair. Conway writes movingly of informal meetings taking place at Clifford's home after the Congress, when, scarcely able to raise his voice above a weak whisper, he nevertheless participated in discussion and inspired the group. Although the Liberal Thinkers group subsequently met several times at Huxley's home, and although it was never formally dissolved, he and Frederick Pollock, John Tyndall, Leslie Stephen and others felt that, without Clifford, the Congress had lost its leading spirit. It had really been set up as a platform for Clifford and it turned out to be in some measure a memorial to him.

From the beginning of September 1878 William's health deteriorated rapidly. He was moved from the family home in Colville Road to lodgings in Old Quebec Street, perhaps to be away from the children, perhaps to be near to the Pollocks who lived nearby in Portman Square. One day, when Moncure Conway and his wife were visiting him there, they met Thomas Huxley who was just leaving. His face was clouded with despair as he said, 'The finest scientific mind born in England for fifty years is dying in that house.' Conway wrote:

> On entering we found Clifford in his armchair, serene and full of his habitual humour, his wife trying to smile. He had just been writing and showing her a skit after the style of Lucian which he repeated to us. It represented Christ and his disciples strolling through Hyde Park and being subjected to questions by suspicious policemen. The talk between St Peter and a policeman was exquisitely comical . . . . He was never more like Socrates than in these last conversations.[24]

Conway reported that, even when it was heartbreakingly clear that death was inevitable, Clifford retained his serenity, and the sparkles of humour never left his eyes. He protected his friends from despairing for him; somehow managing to make them laugh with him.

Frederick Pollock, in his introduction to *Lectures and Essays,* sets out with great tenderness the events and achievements of William's life. He does not neglect the personal portrait, and we read of his love for parties and children and practical jokes and fairy tales. Pollock shows us William with his 'laughter free and clear like a child's and his 'inexhaustible store of merriment'. He writes that it was not possible to be depressed in his company because the charm of his personality banished the fatefully justified anxieties that his friends suffered for him. Clifford's delightful, childlike demeanour came with a thread of uncompromising steel running through it. He was totally intolerant of insincerity in thought, word or deed. He once wrote, in a letter to Pollock, that 'A question of right or wrong knows neither time, place nor expediency.' The holder of such views is committed forever to intellectual conflict. During the whole of the seven years he had spent in London he had thrown himself into his academic work and into the battle against what he called the 'medicine-men' of religious hypocrisy and superstition. This had contributed to his total physical decline and William was finally diagnosed as terminally ill.

Cures for tuberculosis had generally been sought in the warmth of the Mediterranean, in the dry sunshine of Spain and Algeria, or in sanatoria in the clear, cold air of mountain resorts, but there remained one other possibility. As early as 1840 the *Invalid's Guide to Madeira* had been published, and doctors writing in *The Lancet* had described that island as 'a stepping stone for consumptive patients on the road to a more vigorous atmosphere'. Leslie Stephen again showed the practical side of his friendship and, even though William seemed too weak to attempt the journey, he got up a subscription to pay William and Lucy's passages to Madeira. Lucy later told Frederick Maitland that Leslie Stephen and his second wife Julia came to take leave of them on the night before they left on the fateful voyage. She said, 'I shall never forget them. They both looked tall and grave and thin, as if they remembered a world of sorrow, and understood ours, and were half ashamed of the happiness they had discovered for themselves.'[25] It must have been a most poignant parting. Leslie Stephen had been married to Julia for less than a year. Both of them had lost their first partners. Julia and her husband Herbert Duckworth's lives mirrored William and Lucy's. Duckworth had died in 1870 when they had been married for only three years, leaving Julia with three young children.

On 10 January, 1879, William and Lucy again left their two young daughters in the care of others and embarked at Dartmouth on the *Balmoral Castle* for a desperate last-chance voyage to Funchal.

Notes

1. Edward Clodd, *Memories*, Chapman and Hall 1916, p. 37.
2. Gillian Avery and Angela Bell, *Heroes and Heroines in Children's Stories, 1780-1900*, Hodder and Stoughton, 1965, p. 134.
3. Moncure Daniel Conway, *Autobiography, Memories and Experiences*, Cassell 1904. Vol.2, p. 351.
4. Mrs W.K. Clifford, 'Personal Recollections of George Eliot', *Bookman*, October, 1927. pp.1-3.
5. *The George Eliot Letters*, G.S. Haight (ed.), Yale University Press, 1956, Vol VI, p. 21.
6. ibid, Vol VII, p.12.
7. Frederick Pollock, *Introduction* to W.K. Clifford's *Lectures and Essays*, p. 21.
8. Mrs W.K. Clifford, 'A Remembrance of George Eliot', *Nineteenth Century Review*, July 1913, pp.109-118.
9. ALS, (one of thirteen) Valehouse Collection, and G.S. Layard, *Mrs Lynn Linton: Her Life, Letters and Opinions*, Methuen, 1901.
10. As note 3, p.354.
11. Valehouse Collection.
12. ALS, Valehouse Collection.
13. W.K.C's letters in Valehouse Collection.
14. ALS, 12 April, 1876, Valehouse Collection.
15. Mrs W.K. Clifford, *Their Summer Day*, published in *Quiver* as 'by the author of *The Troubles of Chatty and Molly* and *The Dingy House at Kensington*'.
16. Frederick Pollock, *Introduction* to W.K. Clifford's *Lectures and Essays*.

17. F.W. Hirst (ed.), *Early Life and Letters of John Morley*, Macmillan 1927, Vol 2, p. 129.
18. ALS, dated 20 November, 1875, Valehouse Collection.
19. ALS, 1875, Valehouse Collection.
20. As note 19.
21. As note 5, Vol. VI, p. 406.
22. As note 19.
23. ALS (one of six) Valehouse Collection.
24. Moncure Daniel Conway, *Autobiography, Memories and Experiences*. Cassell 1904, p. 361.
25. F.W. Maitland, *Life and Letters of Leslie Stephen,* Duckworth 1906, p. 324.

William's Death in Madeira
1879

I am not and grieve not.
(From Clifford's own epitaph)

Records in Madeira show that six passengers from England disembarked from the *Balmoral Castle* on 15 January, 1879. They appear on passenger lists as Snr and Snra Clifford, Colonel Watts, Snr Clifford, Snr Freeman and Snr Collier. Since they sailed from Dartmouth it is likely that William and Lucy had been visiting the family at Exeter and that William's half-brother travelled with the party. Nothing is known of Colonel Watts and Mr Freeman and they may have been independent travellers. The Hon. John Collier was the son of the first Baron Monkswell. He had met Clifford, who was just five years older than himself, through Thomas Huxley, who had commissioned him to paint William's portrait. John Collier's relationship with Thomas Huxley was curious in that he married two of his daughters. His first wife was Huxley' older daughter, Marion. When she died early in their marriage he proposed to her sister Ethel. At that time, 1889, this was actually against the law in England since the Deceased Wife's Sister Act had not then been passed. Their marriage, in Oslo, caused a furore and split the family, but Thomas Huxley gave the couple his blessing. John Collier had probably begun the portrait of William in 1876 but did not finish it till 1878. During those years, as William became more and more ill, the two had become friends. Collier gave the portrait as a gift to Lucy, and in her will she left it to the Royal Society, where it now hangs. He later painted a replica now held by the National Portrait Gallery. In *The Dictionary of National Biography,* Leslie Stephen describes Collier as Clifford's 'intimate friend', and it was a most noble act of friendship when Collier – then just newly engaged to Marion Huxley – packed his bags and sailed for Madeira to help support William and Lucy on the difficult voyage. Collier took with him his oils, brushes and sketchbook. In spite of the tragic elements of the trip, there would be scenes to paint and the clear atmosphere of Madeira would have been attractive to any artist. The four passengers from England all registered into the Miles Carmo Hotel in Rua do Carmo, Funchal. At that time this was one of the most popular hotels near to the harbour. Only later did it became known as Reid's Old Hotel when the owner built the now famous 'new' hotel on high ground to the west of Funchal.

Funchal at that time did not have a fully developed harbour. The extension of the jetty to the Loo Rock – the Island of Nossa Senhora da Conceiçâo – was begun in 1843, but destroyed by high seas and only completed in 1892. By 1923

William Clifford working at his (rather untidy) desk, by John Collier, c1867. Collier gave Lucy the portrait, and she bequeathed it to the Royal Society.

it was further extended and now forms a safe and attractive haven for shipping. In January, 1879, when the *Balmoral Castle,* under Captain Jones' command, steamed into Funchal Harbour, passengers would have had to disembark into small boats to be rowed ashore. They were met by hotel porters who carried luggage and passengers to the hotels. They were prepared for the arrival of invalids, and would have had on hand some of the *carro de bois* or bullock sledges, that are unique to the island. They had been developed by an ingenious Englishman who, in 1848, had adapted the conventional rough wine-cask sledge into a comfortable form of transport through the uneven steep streets for his semi-invalid wife. They were in everyday use till the 1920s. At the harbour there would also have been some of the old-fashioned hammocks slung on long poles and carried on the shoulders of two sturdy Madeiran peasants. William Clifford was most probably carried the few hundred yards from the port to the hotel entrance by this means.

At that time of the year the temperature would have been between 60 and 70 degrees Fahrenheit, and in that soft climate William seems to have been comfortable and even to have rallied a little. Collier found time for at least two oil sketches painted in the hotel garden. One of them was to prove particularly poignant and very significant. The paintings are not dated but they must have been done between 16 and 23 January since Collier returned to England on the

A postcard of the Miles Carmo Hotel in Funchal as it was when William and Lucy were there. Redevelopment has recently claimed the gardens. The building remains although it is no longer an hotel.

24th. One of them is a charming view of the shady pergola outside the hotel with, in the distance, the mountains that dominate the island. For the second sketch, Collier chose to sit on a balcony jutting out at right angles from the first floor of the hotel. He seems to have nearly completed the oil sketch when William popped his head out of his bedroom window. Collier took up his brush and painted William's head into the picture and thus we have the last ever image of this remarkable man. Today the garden has been taken for redevelopment and only a single palm tree remains. Recent buildings hide the view of the mountains and the former Miles Hotel now houses an embroidery factory and an architect's studio. Collier could never have guessed that the sketch, which passed through Lucy's daughter Ethel's hands and on through the family, was to provide the vital, indeed the only, clue to enable the author in 1995 to track down, first the hotel, and then the rooms occupied by the Cliffords.

More details about those last weeks come from an unusual source. Living in Madeira at that time was a well-known educational figure with a most extraordinary history who knew of, and greatly admired, Clifford's work. This was William Cory the historian, teacher and poet. A brilliant Cambridge prize-winning scholar and classics master at Eton, he was forced, in 1872 at the age of forty-five, to leave his post because some of his friendships with pupils were considered indiscreet. He changed his surname from Johnson to Cory, did some teaching in

Oil sketch of the hotel, by John Collier, with William Clifford (just visible) looking out of the top left-hand window. It has not been possible to reproduce this delightful sketch in colour. Although more recent buildings now surround it, in 1995 it was still possible to stand in the garden and see this view.

Cambridge and London and, in 1877, moved to Madeira for health reasons. He was fifty-six. At this stage, one of his former private pupils – a twenty-year-old girl – fell deeply in love with him, and, in spite of family objections, they were married. This unlikely liaison proved to be a happy one. They were expecting their first child and living in a beautiful Quinta on the mountain slopes outside Funchal when the Cliffords arrived. Alerted by their mutual friend Frederick Pollock, Cory invited them to stay at his home, but Clifford was far too ill to accept the invitation. William Cory did, however, visit him before he died. Cory may not be remembered by many, but the words and the music of the famous *Eton Boating Song* are, and he composed them for the Fourth of July Festivities at Eton College in 1863. His translation from Callimachus, published in 1845, is also well known:

> They told me, Heraclitus, they told me you were dead,
> They brought me bitter news to hear and bitter tears to shed.
> I wept, as I remembered, how often you and I
> Had tired the sun with talking and sent him down the sky.
> And now that you are lying, my dear old Carian guest,
> A handful of grey ashes, long, long ago at rest,
> Still are thy pleasant voices, thy nightingales, awake;
> For Death he taketh all away, but them he cannot take.

Cory wrote to Georgina Pollock on 4 March 1879 and gave us this touching description of William Clifford's last days:

> I heard last night, March 3rd, that Clifford had just died, having been in a dying state for about four days. I hear today that he had no pain .... The last time I saw Clifford he could not talk, and she (Lucy) told me to speak to him louder than before. She seemed to approve of my telling him little things that were meant to be gay and amusing. The time before he had talked to me with point and vivacity about Manning at the Metaphysical, and he listened with interest to what I had told him of the biography of Shadworth Hodgson. Feb 14th we had a very gay ball at his hotel. I had asked beforehand whether it would disturb him. She [Lucy] said that he would like to hear the music and to see my little girl [Cory's young wife] in her ball dress; and on the evening itself she came into the ball-room when we arrived, first of all the guests, and gaily brought us to the sick room where he was lying on the bed, with two candles on a dwarf table by his side. Caroline was struck, and one may say charmed, by the pleasure he showed in greeting her; and when she was a little way off, talking to Mrs Clifford, I saw that although he tried to listen to me, his eyes were fixed on a bright head and a pale blue dress, his last bit of sweet girlish pleasantness. It was as if one had put a flower on his counterpane .... One day he said, when I asked whether the flies bothered him, that he did not so much mind a fly's touch, but he did resent a fly's spoiling his focus by crossing the line of sight. One day I had been reading something about the blind girl Mélanie de Salignac, who was perfect in geometry, and I asked him whether the Lancaster Professor Sanderson was born blind. He had never heard of him. I said that the writer about blind mathematicians said that as the blind perceive lines, angles, curves, etc. by touch, so deaf mutes think in visual images of words (others in sounds). He said that he himself thought not in sounds, but in visual images of words. This seemed to me most interesting. I suppose his eyesight was altogether much livelier than that of other people; perhaps the great chess players have a similar superiority, see twelve sets of chessmen in the head as the geometer sees the most complex crystal or moon-spin. It awed me a bit to sit with a man whose main thoughts were absolutely incommunicable. The fascination lies in that sort of simplicity which

we call childlike, or that grace which we call birdlike. Neoptolemus has visited Philoctetes. I have seen an ingenuous man and had just a glimpse of an edge of a pellucid mind which I cannot measure. I had reckoned on going with my young guest Fred Lees to the funeral in the cemetery, which is clean and beautiful. I do not wish to see a coffin put on board a boat and swung into a ship; perhaps the friends want to meet over a philosophical grave, but I think some philosophers would rather be under the nearest flowers.[1]

On 21 March Cory wrote to another friend:

Clifford I only met twice, three weeks ago, just as his health broke. He was then asking me how Luther went to work, how he came to succeed . . . . He enjoyed seeing with a rare strange power of sight; his eyes spoke when his voice was abated. I fancy he had constant perceptions of space and points and lines which were incommunicable. I have seen very little of men of fine mind. Of his mind I could just get a glimpse. I used to talk with him quietly but gaily, and it amused him . . . . She [Mrs Cory] was charmed with his kind ingenuous look and friendly hand. She wept for him, and for the last hour before sleep mused on widowhood, her own doom; and just her little touches were of more value to Mrs Clifford in grief than all my elaborate hours of talk. But I was of some little use, directing the widow's thoughts from that insatiable brooding worship of the dead man . . . . He was fond of music, children, birds. He was very affectionate with friends. When very near to death he wrote his last letter, just to express affection to Fred Pollock.

During the time that William and Lucy were in Madeira they were not forgotten by friends at home. One of the dearest of them, Leslie Stephen, 'the most loveable of men' as James Russell Lowell described him, wrote to William on 6 February, 1879: 'We were very much pleased to hear that you had got through your voyage tolerably well and were finding yourself the better for Madeira . . . . It seems to me an age since you went off that frosty morning and my pleasant visits to Quebec Street came to an end.' He goes on to tell of his latest visit to the Alps and gives a critical review of his thoughts about Huxley's article on Hume and of the latest meeting of the Metaphysical Society. He writes of the book he is planning saying 'I should much like to talk to you about some of the points which I want to state; but alas the telephone is not yet arranged to Madeira – or you would be bothered with my remarks pretty often. I hope we may discuss it hereafter, unless I get so disgusted that the whole thing goes into the fire.'[2] Leslie Stephen must have known that he would not see his friend again: indeed, Clifford lasted just a few more weeks.

John Collier returned to Southampton on 24 January, leaving William a little more comfortable in the softer climate of Madeira. But death was inevitable. Pollock records that Clifford was cheerful, clear-headed and interested in the daily news till the day he died. Knowing that the end was near, he had managed to write some last messages. He left detailed instructions about his academic works, which he considered the only important thing of his life and he wrote this last and most moving letter to Fred Pollock:

> My dearest Fred, I am told this must be written before I get dazed, as I may not get clear enough again. No words can say what a friend you have been to me, I who have had so many and so good ones, must always count you the best and the truest. I have nothing to leave you but my children, the education of which I entrust to you and Huxley. I want them brought up without any knowledge of theological – that last syllable (!) – hypotheses at all, but if the theory should teach anything of the sort it should be set aside with the simple argument suited to the form it is presented in – for the rest of their education, which I should probably spoil, I must trust you . . . but if my most beloved and devoted wife Lucy, who is the best loved that ever lived, should survive the shock of my death, she will of course take care of such things herself. Give my love to Sir Frederick and my Lady, to my dearest Georgie and your kids, to the Walters and Jack and to all that have it. Yours always, Willi.[3]

Even though his strength was fading fast, William could not resist that last stab at what he perceived as the inaccuracy of the word 'theological'.

In his last hours, perhaps echoing Epicurean philosophy, he composed his epitaph. The original script, in shaky capital letters, show how much this final effort cost him:

>> I was not, and was conceived:
>> I loved and did a little work.
>> I am not, and grieve not.

By the time he came to write something for his two daughters he could manage only the briefest line to each of them.

William had asked that his body should be returned to England and buried on high ground and near a tree. Lucy chose Highgate Cemetery and Fred Pollock purchased a burial plot there. There were some who publicly objected to the faithless inscription which William had composed for his tombstone, but he had inspired such respect during his life that all such objections were set aside.

To Cory, and perhaps to others, the returning of a body to England seemed unnecessary when a pleasant English cemetery was available in Funchal. Madeira was of course a predominantly Roman Catholic island, but there is an English Church there, which was the natural centre of the religious and social life of the non-Catholic members of the community and visitors to it. However, neither William nor Lucy made contact with it. Before 1765, burial of non-believers was not allowed on the island. Their bodies were buried at sea or simply thrown from the cliffs. Those barbaric days were of the past, but Lucy would have known that it would be inappropriate to bury her husband so far from the people who loved him and recognised his genius. However, there would have been bureaucratic and practical difficulties to overcome, not the least of them being the problem of finding a suitable ship for the return of the body. Lucy needed influence and local help to carry out her plan and it is possible that it came through one of the most renowned of all British scientists of the time, Sir William Thomson.

Details of William Clifford's death on March 4 had been received in London in time for an obituary to appear in *Nature* on March 7. Marconi did not send his famous radio signal across the Atlantic till 1901, but earlier communication was possible through undersea cables. These had been laid from Brazil to Madeira in 1865 and later extended to Portugal from where there were overland links to England. The Trans-Atlantic cable-laying had been under the supervision of William Thomson (1824-1907), who became Lord Kelvin. He and William Clifford had met at the Royal Society and British Association meetings and had mutual friends in James Clerk Maxwell and Thomas Huxley. The curious coincidence was that William Thomson had family links in Madeira. In 1873 Thomson had been held up for sixteen days in Funchal Harbour for repair work to his cable-laying equipment. He was at that time fifty-five years of age. He was a most brilliant scientist – one of the best-known academics of all time; he was wealthy, and he was a widower. Charles R. Blandy, the powerful and very wealthy Madeiran ship-owner and businessman, had two attractive daughters and Sir William was a frequent visitor at his family home. With the cable repair mission completed, Sir William departed. However, a romantic liaison had occurred, and the next year he returned to Madeira and married one of the daughters, Frances Anna Blandy. It is pretty certain that he was not in Madeira during William's dying days, but it seems likely that he intervened, perhaps getting Gladstone's support in England and Blandy's practical help in Funchal, to enable William's body to be returned to England on a gunboat that had been diverted to Funchal on its return voyage from the Zulu Wars. For Lucy, the island of hope had become the island of despair. She remained in Funchal till March 15 before returning to Plymouth on the *Balmoral Castle*, again with Captain Jones in charge, to face life in London without her husband.

Earlier, on Sunday, 13 March, a Memorial Service arranged by Moncure Conway took place at South Place Chapel. At it, Conway delivered a commemorative discourse as a tribute to William and he also contributed an article about him for the monthly journal *Modern Thought*. Later, as he stood heartbroken at William Clifford's graveside, he likened William's early death to a 'calamity that had befallen the Round Table of liberal thought', and continued the analogy by quoting, from the *Morte d'Arthur*, the words spoken of Lancelot:

> And now I dare say that Sir Lancelot, ther thou lyest, thou that were never matched of none earthly knyght's hands. And thou were the curtiest knyght that ever bare sheilde. And thou were the truest freende to thy lover that ever bestrode horse; and thou were the truest lover of a synful man that ever loved woman, And thou were the kindest man that ever struck with swerde. And thou were the goodliest person that ever came among preece of knyghts. And thou were the meekest and gentillest that ever sate in hal among ladies. And thou were the sternest knyght to thy mortal foe that ever put spere in the rest.[4]

Sir James Crichton Browne, the eminent psychologist, wrote that the

saddest thing about Clifford's early death was that: the world was left with his speculative ingenuity in an immature stage . . . who can say that the pendulum of belief might not have swung back to a point very near that which it reached in his earlier days when as a student of Aquinas he was dominated by an intense and living faith . . . and supported Catholic doctrine with singular scientific ingenuity.[5]

The idea that Clifford might join the Roman Catholic Church had been aired in the press during his lifetime and had elicited this amusing note from him:

To the Editor of the Pall Mall Gazette. I was fairly astonished to see in your columns today a report that I had joined the Roman Catholic Church. I should have been amused at its incongruity, but that the report amounts to a serious charge against me unless I have ceased to be responsible. It is true that I have been somewhat unwell of late, but I am assured by Dr Andrew Clark that my indisposition had not yet taken the form of mental derangement. I remain, Yours etc, W.K. Clifford.

Certainly, William's vigorous agnostic beliefs had offended many, but some that opposed him nevertheless respected him. W.H. Thompson, Professor of Greek and Master of Trinity College at the time when Clifford was elected Fellow, had declined to sign the public testimonial published as Clifford was dying. However, one year later he sent Pollock a substantial financial contribution with instructions that it be given to Lucy anonymously from, 'One who had as much admiration for her late husband's talents as he had disapproval of his philosophical opinions.'

The academic world had lost one of its brightest stars. Obituaries and glowing tributes appeared in newspapers and journals and his friends and colleagues set about collecting and preparing for publication his lectures, essays and other writings. Frederick Pollock's words, describing the days after it was clear that Clifford could not recover, provide a most moving valedictory:

From that day the fight was a losing one, though fought with such tenacity of life that sometimes the inevitable end seemed as if it might yet be put far off. Clifford's patience, cheerfulness, unselfishness, and continued interest in his friends and in what was going on in the world were unbroken, and unabated through all that heavy time. Far be it from me, as it was far from him, to grudge to any man or woman the hope or comfort that may be found in sincere expectation of a better life to come. But let this be set down and remembered, plainly and openly, for the instruction and rebuke of those who fancy that their dogmas have a monopoly of happiness, and will not face the fact that there are true men – ay, and women, to whom the dignity of manhood and the fellowship of this life . . . are sufficient to bear the weight of both life and death. Here was a man who utterly dismissed from his thoughts, as being unprofitable or worse, all speculations on a future or unseen world; a man to whom life was holy and precious, a thing not to be despised, but used with joyfulness; a soul full of life and light, ever longing for activity, ever counting what was achieved as not worthy to be reckoned in comparison with what was left to do. And this is the witness of his ending, that as never man loved life more, so never man feared death less. He fulfilled well and truly that great saying of Spinoza, often in his mind and on his lips: *Homo liber de nulla re minus quam de morte cogitat.*[6]

Notes

1. This and the quotation following it are from Willam Cory, *Letters and Journal* 1823-1892, Oxford 1897 (published for the 25 subscribers), pp.441-443.
2. F.W. Maitland, *Life and Letters of Leslie Stephen,* Duckworth and Co. 1905, p. 333.
3. ALS, Valehouse Collection.
4. M.D. Conway, *Autobiography, Memories and Experiences*, Cassell 1904.
5. Sir James Crichton Browne, *Victorian Jottings*, Etchells and Macdonald, London, 1926.
6. Frederick Pollock, Vol 1, *Lectures and Essays*, W.K. Clifford, Macmillan 1879, Introduction.

Part Two
LUCY ALONE 1879-1929

Beginning Again

The following announcement had appeared in the Times on 17 March, 1879.

> The Friends of Professor Clifford, who was compelled by ill health to relinquish active work and reside in Madeira, were anxious to present him with a substantial Testimonial in public recognition of his great scientific and literary attainments. At a Meeting held at the Royal Institution, Albermarle Street, on Friday 31 January with William Spottiswood, Esq. President of the Royal Society, in the Chair, it was resolved that a Fund should be raised for the above-mentioned purpose, and that the sums received would be placed in the hands of Trustees for the benefit of Professor Clifford and his family.

An amendment was added:

> Professor Clifford died at Madeira on March 3rd. The Executive Committee have now to announce that it is intended to proceed with the raising of the Fund for the benefit of his Widow and Children who have no other provision.

There followed a list of the names of the General Committee and over two hundred supporters who had already subscribed. This glittering array forms a directory of the great names in academic and cultural life of the time, and stands by itself as a contemporary indicator of the great regard in which William Clifford was held. The total contributions amounted to £2,300. William Spottiswood later wrote that while the Fund was 'eminently successful in the number of subscriptions, yet the actual contributions were (as indeed was expected) small. The sum arising can serve only as a nucleus for accretions from other sources.' The annual salary of a Professor in the University of London at that time would have been in the region of £300. The fund money was invested under the trusteeship of Sir John Lubbock and William Spottiswood, and the interest from it, £90 a year, formed the backbone of Lucy's income in her widowhood.

George Eliot had responded immediately to news of William's death. On March 7 1879 she noted in her diary: 'Sent subscription of £10 to Clifford Fund.' she wrote to Lucy as soon as she arrived back from Madeira, offering to visit her and inviting her to tea. In her letter, George Eliot linked her loss of George Lewes four months earlier with Lucy's loss of William and shared the sorrow of it with these words:

> I understand it all .... There is but one refuge – the having much to do. You have the mother's duties. Not that these can yet make your life other than a burden to be patiently borne. Nothing can, except the gradual adaptation

of your soul to the new conditions . . . . It is among my most cherished memories that I knew your husband and from the first delighted in him . . . . All blessing – and even the sorrow that is a form of love has a heart of blessing – is tenderly wished for you.[1]

Lucy later gave the letter to Matilde Blind who used part of it in her 1883 biography of George Eliot.[2] When she went to the Priory for tea on 25 April, 1879, George Eliot eagerly asked her: 'Do the children help? Does it make any difference?'[3] When they were alone one afternoon, Lucy reports:

> Mrs Lewes made a little sign that took me from the arm-chair on her right to a grey, cushioned footstool by her side. She took off my hat, and so we sat, she talking and I listening. Now and then she put her wonderful hands on my hair, they sent a thrill through me – the memory does; even yet . . . . She had a wonderful personality, and with the exception of my husband, greater magnetism than anyone I ever knew. Something indefinable looked out of her grave eyes and lurked in the fleeting smile.[4]

Lucy also wrote about another meeting:

> The last time I saw her she was not alone – it was just before her marriage to Mr Cross. Leslie Stephen was there, and she talked chiefly to him, but she held my hand in hers the while – she had the most soothing touch of any woman I ever knew; and to feel her hand in yours was to be sensible of all the troubled ways in your life peacefully subsiding . . . . We asked if we should see her again before she left for her Surrey cottage. A happy smile that vaguely puzzled us came to her face while she answered, 'Oh, yes, I dare say you'll see me – or you'll hear of me.' We did – we all heard of her marriage to Mr Cross.[5]

The wedding took place in May 1880, but George Eliot fell ill in October and by the end of that year she was dead.

After her return from Madeira, Lucy recuperated for some weeks with her mother in Connaught Square. She took chloral to help her to sleep and experienced some vivid dreams, which she wrote about on 17 May, 1879. She felt it was important that William had told her in the dreams that she would not be able to see him again because she was amongst the 'middle shadows'. As she recovered, she began to establish for herself the philosophy that would sustain her through the next fifty years. She wrote to Frederick Pollock:

> I am not so much to be pitied. I had nearly six years of perfect companionship (for we saw each other almost daily for eighteen months or so before we were married) and found reason every day to love and reverence him more – and I find it still; and see more and more (tho' it has been my strange good fortune to know the best and greatest of men) that there was and is no one so perfect or so great (not even you, my dear old ugly Fred). There are not

so many women after all that have this blessedness – especially women that have, as I have, a horrible power of keeping their critical faculties unweakened by their affection.[6]

She wrote often to the Pollocks remembering William: 'How simple and how happy he was and what a ringing laugh he had with a little shout at the end. There was such a wonderful light and life and brightness about him.' She told how William used often to say 'be free' at the end of his letters; 'He said it was an old form and much better than goodbye which was full of superstition.' Lucy's love for her husband comes through in these letters; she became quite fanciful and 'wild' as she grieved for him in her early widowhood. She wrote:

> Supposing there is after-consciousness . . . could it be possible for many forms of intellect and beauty to take refuge in one physical frame until they made up a perfect whole, worthy of standing alone; so that Willi represented the former consciousness of many and is after all living still or carrying on in some other world what is just going on in this – the survival of the fittest.[7]

Lucy's distress at losing William caused her to explore these ideas to comfort herself for her loss. She felt that if the strongest and greatest survive, and are 'grouped after the molecule and atom theory' then it could account for William's 'many sidedness, his many forms of greatness, imperfect only from accident or physical restraint'. She knew that William would not have agreed with her thoughts, and neither would Huxley or Pollock. She later excused herself for her fancies, but they never entirely left her.

Of course she was not without friends in London. Some were powerful, some were wealthy, and all were ready to help. Eventually Lucy was well enough to take up her journalistic work again, but finance was her most pressing problem. George Eliot supported Thomas Huxley and William Spottiswood in their approach to the Royal Literary Fund for an award for Lucy. In Lucy's own letter of application she wrote:

> Our income, which depended solely on my husband's personal exertions and therefore on his health, was gradually diminished by his long illness and at his death all resources were absolutely exhausted. My father-in-law died at a time when my husband's health and prospects appeared to us to be excellent, and property which might have come to us was left entirely to a young family by a second wife. My sole income at present is £90 a year derived from a Testimonial Fund raised by public subscription to my husband's memory. Later I hope to supplement this a little by my own work but as yet this has been impossible.[8]

A grant of £200 was awarded, but she still needed to add to her day-to-day income and she became a literary correspondent for *The Standard*. Her daughters later remembered the house always being full of stacks of books for her to review. Norman MacColl, who had been a friend before William's death, was very supportive. He was, at that time, editor of the *Athenaeum,* and he invited Lucy to contribute to that journal's gossip column. Later, his extraordinarily generous legacy of £2,000 to Lucy's daughters provided a dowry for Ethel in 1905 and financial security for Margaret, who never married. Lucy's natural talent for friendship was a tremendous asset. At her congenial Sunday tea parties her guests exchanged news and views of books, the theatre and current literary gossip, which kept Lucy at the heart of London cultural life. Established writers and influential editors together with young hopefuls filled her little drawing room. Frederick Pollock and his wife, now with their daughter Alice and son John, were regular visitors. John Pollock, recalls 'Aunt Lucy', as he and his sister called her, as a woman 'as remarkable for character as for talent'.[9] He writes that from his childhood till Lucy Clifford's death 'there was never a time when she did not occupy a significant place in my mind'. He knew her as 'a successful literary critic and novelist and as a charming, witty and popular hostess'. He added that 'if the conversation ever flagged, she could always rely on the writer Violet Hunt's vitriolic tongue to start a fresh hare'.

Among the regular visitors to her home he remembered Ford Madox Hueffer, F. Anstey of *Vice Versa* fame, Elizabeth Robins the novelist and actress, George Street, Gerald Duckworth, George and Frederick Macmillan, Somerset Maugham, Norman McColl, Henry James, and, in the early days, Rudyard Kipling. Lucy was Sir Frederick Pollock's – he had inherited the title in 1888 – longest-standing friend and his country house at Hindhead was a second home for her and her daughters. She set several of her stories in that thinly populated heath-and-heather countryside where the air was pure and healthy. John Tyndall, another of William's eminent academic colleagues, had started a fashion by building a second home near Haslemere in 1883, and a colony of writers, philosophers and scientists of the late Victorian and Edwardian age established itself in that part of south-west Surrey. Tennyson, Mrs Humphry Ward, Bernard Shaw, Sir Arthur Conan Doyle, Grant Allen, George Eliot and Frederic Harrison were among the many who had homes there at one time. The Tyndalls and Pollocks were generous hosts and John Pollock paints a picture of all year round activities and a constant stream of visitors. There was cricket, tennis, swimming, skating and sledging, walking and lunching, and all the time stimulating talk and discussion. Supported by this close circle of friends, Lucy began writing novels. Her many books are discussed in a later chapter, and, by 1885, when her book *Mrs Keith's Crime* hit the headlines, she was a regular contributor to various newspapers and journals and had three children's books to her name.

Notes

1. G.S. Haight (ed.), *The George Eliot Letters*, OUP 1954-56, Vol. vii., p. 215.
2. Matilde Blind, *George Eliot*, W.H. Allen and Co. 1883, p. 215
3. ibid.
4. Mrs W.K. Clifford, A Remembrance of George Eliot, *Nineteenth Century Review*, July 1913. p. 109.
5. ibid.
6. Valehouse Collection.
7. ibid.
8. British Library, Royal Literary Fund Papers. Case File 2086.
9. Sir John Pollock, *Time's Chariot*, John Murray 1950, p. 80.

Friendship with Rudyard Kipling

Lucy's relationship with Kipling began when she was an established writer and successful hostess in London. A friend had sent her from India a book of short stories titled *Plain Tales from the Hills.* She recognised the high quality of the writing and, in 1889, when its author, the young Rudyard Kipling, arrived in London, she befriended him and introduced him to Macmillan to get his work published. He was so often at Lucy's home that her friend, Marie Belloc Lowndes, wrote that he almost seemed to have taken the place in her heart of the son she and William had longed for.[1] Lucy pulled strings to get him elected to the Savile Club – a Mecca for young writers. She told of him hurrying round to her house in the night, banging on her door and demanding to be let in to read aloud to her a story he had just written. It was *Without Benefit of Clergy,* and, when she encouraged him to go ahead and publish, he gave her the manuscript as a keepsake. In 1921 Lucy sold the Kipling manuscript to James Russell Lowell's sister, Amy.[2] Later, in 1921, she wrote to Scribner in New York asking if he knew a private collector who would purchase books and letters she had from Kipling. She wrote 'I should hate to sell, but I have an unmarried daughter to provide for and the grey covered Kipling books inscribed with his name and mine would be worth £50 at least.'[3] 'Rudd', as Lucy called him, enjoyed being with young Ethel and Margaret, who were then thirteen and twelve. When they had a new black kitten he 'stood godfather' to it and solemnly christened it 'Skuttles' and presented it with a collar on which he had had engraved 'My name is Skuttles. I live at 26 Colville Road. Please take me home'.

    A delightful remembrance of Kipling – a striking portrait – resulted from his friendship with Lucy Clifford. It hangs in the east bedroom at Rudyard Kipling's home, Bateman's, now a National Trust property. It is a head and shoulders painting of a youthful Kipling in a white Indian tunic. It is strangely compelling: the mouth is partly concealed by a moustache and the eyes, though kindly, are shrewdly observant. Both it, and the magnificent 1900 portrait which hangs on the staircase at Bateman's, were painted by the Hon. John Collier. Lucy had asked him to paint her new young friend freshly arrived from India. Collier showed this first portrait of Kipling at the New Gallery in 1891. He remained a close friend to Lucy after William's death and he gave her the portrait. Lucy displayed in her home for twenty years. Another fascinating image is the previously unpublished pencil sketch, done by his father, which Rudyard gave Lucy early in their friendship. It shows Kipling at work, his eyes on the page with smoke from his pipe tracing the word '*Finis'*. It captures the intensity of the writer at work. He has inscribed it: '*Rudyard Kipling to Lucy Clifford. June '90.*'

When he arrived in London in October 1889 he took rooms in Villiers Street, which runs between the Embankment and the Strand, and Lucy visited him there. She never wasted the fruits of her observation, and in her book *Miss Fingal* she describes the Villiers Street lodging, and makes one of her characters a pipe-smoker so that he may 'use' Kipling's pipe-rack. Kipling was particularly fond of Lucy's younger daughter Margaret. Three of his charming letters to her are published in full in *The Letters of Rudyard Kipling*.[5] In 1890, Margaret was away at school in Clifton and feeling homesick. She was thirteen and Rudyard twenty-five. He wrote jolly, teasing, fanciful letters with amusing drawings in the margins to cheer her up. To encourage her to learn to play tennis he offered to buy her a new racquet and it is clear that he very much enjoyed his easy relationship with Lucy and her girls. In one of these letters he writes:

Revered Turks,
   Your Mummy is kite well acause I has just seen her an tomoworo we goes to Hazlemere to play wif Tennyson [at the home of Sir Frederick Pollock] as soon as I have a real live address I'll write or telegraph it to you, but just you go on writing to Embankment Chambers, Villiers St., Strand and the letters will tumble in somehow. I am nearly broked in two. I have done two books an' I'm dead tired and frabjous an' muzzy about the head. Likewise polumneas and metheoligastical which are serious diseases.

In another, addressed to 'Dear Turks', he mentions, 'Your Mummy has got a play on her brain and I've got one on mine so you can imagine when we both talk at once about our own play, how pleasantly and intelligently we converse!'

The letter following was probably written in 1890, when he was lodging in Villiers Street:

        December 22. Embankment Chambers, Charing Cross.
Dear Lady,
   This is very offle – but I'm not sorry 'cept that you have chosen a vile bad time for your cantrips. Christmas is not, permit me to remind you, the best season for measles but saith the prophet, any time and every time is good for rest, and rest it is that you most need. I thought I had 'em last night – cause I was awful cold. Then it occurred to me that plowtering from the Mothers through the thaw, was not the best way of getting warm. Then I did not have any more measles. You're well out of today's fog which is solid enough to knock down the piers of Charing Cross Station. The trains are hooting spitefully at each other but all they can do is to let off fog-signals and run wildly up and down the bridge. Don't think much of a horn in a fog myself. Too like a lady with measles – of no use to anyone except the owner. Oh damn – a man from the Detroit Free Press has just come up and I must tend to him. I dine on Boxing day with Sid Colvin (and Pantomime afterwards sez he. 'Sid'sez I 'I'll see you minus your pants before I'll go to a pantomime on the 26th Dec.') . . . . Keep well, don't fash yourself and goodbye till this even when a small note comes over from Ruddy.[6]

Rudyard Kipling by John Collier, who gave it to Lucy in 1891. It hung in her drawing room until 1911 when she gave it to Kipling in exchange for one of his manuscripts.

In complete contrast to these relaxed and friendly notes comes one dated 11 August, 1891. It is tight and formal:

Dear Mrs Clifford. You have my full authorization to dramatize my novel The Light that Failed in any manner and under any title that may seem good to you, upon the terms agreed upon between us.
Sincerely, Rudyard Kipling.[7]

Lucy did not take up this authorisation, but Kipling did later have the story dramatised, and in 1903 it was produced at the Lyric theatre. A film version was made in 1916. The letter may have been couched in formal terms for business

Pencil sketch of Rudyard Kipling, drawn by his father. This must have been one of Rudyard's earliest gifts to Lucy. It is inscribed 'June '90' but may have been drawn at an earlier date.

purposes, but it was written soon after the distressing and dramatic break-up of his friendship with Lucy. Marie Belloc-Lowndes recounts how Rudyard and Lucy were the closest of friends until Lucy, in response to something Rudyard had said, made an overly critical comment about the rather plain-faced Caroline Balestier, not realising that she was the woman he would later marry. Rudyard Kipling reacted violently and broke off the friendship. Lucy was heartbroken. Mrs Belloc Lowndes wrote: 'Not long before she died, she told me that when she came across Mr Kipling, as was inevitable now and then, a sensation of such pain filled her heart that she had at once to leave the room where he happened to be'.[8] Sir John Pollock gives a different account. He writes: 'One day at a tea party they [Lucy and Rudyard] were standing together when he said "Good gracious, isn't that the plainest woman you ever saw?" Mrs Clifford laughingly assented.' Kipling later regretted this shared exchange and broke off his relationship, not only with Lucy, but also with Frederick Pollock and his brother Walter.[9] Whatever the actual details were, it was a grievous loss of friendship to all of them. Later, Ethel, Lady Dilke, gave this description of the manner by which Kipling had broken with Lucy:

> His manners were little better than they had been in India . . . . He seldom answered invitations, and Mrs Clifford was forced into a number of awkward and irritating apologies on his behalf. There was a painful moment in their association when Mrs Clifford . . . suddenly asked him at dinner: 'But Ruddy, aren't you supposed to be dining with – ?' Kipling replied 'Oh yes, but I sent them a wire.' . . . the episode must have lingered unpleasantly in her mind, causing her to warn him later: 'You mustn't do it, or one day you will send one to me'; and so, in the end, he did.[10]

Twenty years after the painful break-up Lucy wrote him a very significant letter. On 15 December, 1911, Kipling replied to it. He wrote on notepaper from The Athenaeum and marked it 'Also Private':

> I have just received your most kind letter. By all means dispose of the mss of 'Without Benefit Of Clergy' as you think best and I will gladly consider the Collier portrait as ample and generous exchange. It is like your kind self to think of it. I understand perfectly how you feel in the matter. I am sorry that Turkey [Lucy's daughter Margaret] has gone in for Theosophy. It is skittles for any colour but specially skittles for white people. I am middle-aged, grey and bald but none the less, gratefully yours, Ruddy.[11]

Probably it was at this time that Lucy Clifford's name, entered as the owner of the portrait in John Collier's 'Sitter's Book', was crossed out and replaced by 'Mrs Kipling senior' who always considered that the portrait was by far the best ever done of her husband.[12] Some of the painful memories of their estrangement must have been healed by this generous gesture.

A long letter from Rudyard to Ethel Clifford, dated 23 September, 1933, indicates that she had taken up the friendship again after her mother's death. She had appealed to Rudyard for advice about the suitability of the climate of South Africa for her son's health. In his solicitous reply from Bateman's he suggested that it was probably not a good idea to engage upon such a hazardous trip without very careful considerations. He ends the letter 'Always with remembrance and affection. Rudyard Kipling.'[13]

A recent study of Kipling [14] casts a new light on the break-up between himself and Lucy. Rudyard's sister Trix held the view that Lucy wanted more from their relationship and spoke of remembering how shocked Rudyard had seemed when Lucy had raised the subject of matrimony. Trix also thought that Lucy had painted Rudyard into Jim Alford, the love-of-the-life of her heroine in her autobiographical novel *A Flash of Summer*. Certainly Rudyard's interest and attention to Ethel and Margaret indicated the strong 'family' aspect of his friendship; but there was a twenty-year age difference between him and Lucy and a much greater difference in personality and attitude to accommodate. However, one can imagine an affectionate comment being misinterpreted in what must have been a finely balanced relationship during Rudyard's early days in London.

In his autobiographical book *Something of Myself*, Kipling writes of his early days in London:

> my small stock-in-trade of books had become known in certain quarters; and there was an evident demand for my stuff. I do not recall that I stirred a hand to help myself. Things happened to me. I went, by invitation, to Mowbray Morris the editor of *Macmillan's Magazine* . . . who took from me an Indian tale and some verses . . . both published in the same number of the Magazine.

He goes on to describe his 'amazing luck' in the easy acceptance of his work in journals and newspapers. He described how, to his pride, he was elected a Member of the Savile Club where he met, among others, Thomas Hardy, Walter Besant and Edmund Gosse. He greatly enjoyed the stimulating and tolerant atmosphere of the club and the short-cut to acquaintance with the most interesting literary personalities that membership of it afforded. It is a shame that he did not spare a word of recognition for Lucy' generosity. Clearly, he would have made his way without her, but he was most fortunate to have had such a warm-hearted protagonist in his early days in London.

Notes

1. Mrs Belloc Lowndes, *The Merry Wives of Westminster*, Macmillan 1946, p. 65.
2. ALS, Amy Lowell, 29 December, 1911, Valehouse Collection.
3. Archive of Charles Scribner's Sons, (CO101) Box 32, Rare Books and Special Collections, Princeton University Library.
4. Lord Birkenhead, *Rudyard Kipling*, Weidenfeld and Nicolson, 1978.
5. Thomas Pinney (ed.), *The Letters of Rudyard Kipling*, University of Iowa Press, 1990, Vol 2 p. 17.
6. ALS (one of four), Valehouse Collection.
7. ibid.
8. As note 1.
9. Sir John Pollock, *Time's Chariot*, John Murray 1950, p. 91.
10. As note 4.
11. ALS, Valehouse Collection.
12. National Portrait Galley Archive. *John Collier's Register*, 1891, entry for Portrait of Rudyard Kipling in White Jacket.
13. ALS, Valehouse Collection.
14. *Kipling*, Andrew Lycett, Wiedenfeld and Nicolson, 1999.

Literary Critic and Hostess

The publisher Sir Frederick Macmillan had known and admired William Clifford and he kept up his friendship with Lucy. He and his wife were regular guests at Chilworth Street and both attended Lucy's funeral. Even in Lucy's later days when she was by no means rich, the writer Charles Morgan noted that 'the company in her little drawing room was among the best, though not the smartest, in London'. Sir Frederick trusted her judgement as one of his readers.[1] He considered Lucy to have a vital quality – 'an intuition for literature, and an incapacity to lie about it whether in flattery or spite' and therein lay her strength. Sir Frederick respected her opinions with an 'affectionate . . . reasoned and almost superstitious regard'. She had brought Kipling to them in 1890, and in 1898 she persuaded Frederick Macmillan to back Maurice Hewlett's historical romance *The Forest Lovers*. It had been turned down by the first reader, but Lucy, who knew Hewlett and had encouraged him, read it and judged that it would be extremely popular. It then went to Lord Morley but he was not in favour of it. However Lucy took Sir Frederick to task, declaring that she was certain of its success and that if he rejected it he was 'throwing away a small fortune'. She made him promise to read the manuscript himself. Lucy won the day; the book was published and Hewlett's reputation was made. It is therefore easy to understand the sincerity of what he wrote in 1901 to Lucy: 'God Bless you. He did bless me when he brought me to your door some fifteen years ago'.[2] After Hewlett's death, when an edition of his letters was published and caused his wife unhappiness, Lucy again endeared herself to his memory. Ignoring the undercurrents and gossip she wrote an appreciation for the *Observer* and earned these words from his wife:

> Darling friend, . . . You were his fairy Godmother for years and he loved you . . . . You gave him his first chance and his last tribute . . . he always trusted you . . . you couldn't have done Maurice a better turn than by writing . . . . Oh! how I bless you . . . you knew nothing of the fight and simply wrote from your own knowledge and criticism . . . . I shall burn all his letters now, and then no one can be foolish and rake up what was never meant for the world to see.[3]

Much later in her life, she again did well for Macmillan. Charles Morgan remembers the 80-year-old Sir Frederick along with 'the great survivors of the Victorian age' still visiting Lucy and being influenced by her opinions.[4] Charles Morgan's first novel: *My Name is Legion* had been a great success.

When, in 1927, his second novel was rejected by Heinemann, he was shattered. He sent the manuscript to Lucy whom he respected because, as he wrote to her, she was:

> an artist whose method is much more straightforward than mine so that I am certain you will not start with the favourable prejudice of some côterie kindred to me; and secondly because, though you yourself are a swift and vigorous narrator, you had an ear for Henry James whose manner was infinitely oblique. Therefore you are the perfect critic for my book.[5]

Lucy again insisted that Sir Frederick read it himself, and so set it on its way – re-titled *Portrait in a Mirror* – to be a best seller and, in 1930, to win the Femina-Vie Heureuse prize. Charles Morgan recorded his gratitude to Lucy in these words:

> You have done one service that I shall never forget in bringing me to Macmillan. I have wished to be published by them since, as a child, I realized that those green-bound volumes in my father's library were immortality's mortal representatives! I am sure that between author and Publisher, long association is everything; disloyalty and impatience are the unforgivable sins – an old-fashioned view. Pray heaven the Macmillans share it; if they don't no one does.[6]

A friend wrote of Lucy that she always 'had some mute inglorious Milton up her sleeve . . . . She was always eager to press the claims and advance the interests of those who won her sympathy.'[7] She had been supportive of Noel Coward in his early days. Much later, in 1927, she wrote to him asking if he would help dramatize a book of hers. He replied to her affectionately saying: 'I think your story *Wild Proxy* is excellent but, much to my regret, I can never work on plots that are not my own.' He was glad that she had been in touch and told her to make sure she came round to his dressing room at the theatre.[8] Hugh Walpole tells how Lucy, knowing how he admired Thomas Hardy, took him to meet the great man. Lucy, Hugh Walpole and Hardy sat in helpless silence while Mrs Hardy dominated the conversation. Walpole recalls that, just before they were to leave, Hardy finally got a chance to speak:

> Mr Walpole, I understand you intend to write novels. 'Yes' I cried, ready to burst into a grand exposition of my fine purpose and noble ambitions. 'Don't' he said, sadly patting my shoulder, and away we went![9]

Even when Lucy's literary success was waning, her Sunday afternoon tea parties continued to flourish. Ezra Pound, when he was in London between 1916 and 1919, added some eccentric colour to number 7 Chilworth Street, Lucy's home from 1899 till her death. Theodora Bosanquet, Henry James's secretary, who admired Mrs Clifford, noted that:

Ezra Pound in his brown velvet coat was sunk into his armchair and would have been extremely interesting and was being very witty, only one can only hear half what the creature says in his languid murmur. His account of a tea-party at Marie Corelli's was full of good things, of which I only caught detached fragments of embroidery.[10]

The portrait painter John Collier remained faithful to Lucy after William's death. He was able to carry off what is surely a precarious test of friendship – he could give the most trenchant criticisms of some of her books without her taking offence.[11] He did some illustrations for two of Lucy's books including a beautifully drawn impression of Mrs Keith for the second edition of *Mrs Keith's Crime*. Besides being a most talented portrait painter he was a freethinking philosopher. In his book, *The Religion of an Artist,* he writes: 'Most people assume that some sort of religion is necessary. I do not see the necessity; some of the best people I have ever known have no religion at all.' There is no doubt that amongst 'some of the best people' would have been his twice-time father-in law, Thomas Huxley, and the friend they both loved – William Clifford. He kept up his friendship with Lucy till her death and, with his wife, was frequently at her Salon.

Lucy's day-to-day life in the early London days of her widowhood was very busy indeed. She had to organize her salon, establish herself as a writer, and care for her children. Her fourteen-year commitment to *The Standard* had been taken up soon after William's death. It was a daily paper and, although most of the regular contributors to it are not named, many of the entries under 'Novels of the Day' and 'Christmas Books' are certainly recognisable as Lucy Clifford's work. Lucy's brother, John Lane, also worked for *The Standard* as society correspondent. He was known as a first-rate journalist and extremely popular colleague. Jack Pollock, himself a contributor to *The Standard*, recalls the extraordinary fact that Lucy never spoke to him or acknowledged him as her brother.[12] This untypical coldness was surely a bitter echo of how much early family dislocations had scarred Lucy.

In the early months of 1881 Lucy visited St Germain-en-Laye, west of Paris, and it was to become a favourite escape from the harsh London winter. She used it as a setting for one of her novels, *The Long Duel,* and, in 1888 and 1889, took Ethel and Margaret there with her to improve their French. For another escape from London she accepted the Stephens' offer of Talland House and spent many months there in 1884 and 1885. Olive Schreiner and Leslie Stephen visited her when she stayed at Clarens, near Montreux. This was a favourite resort for writers, poets and musicians and Lucy spent time there in each of the last four years of the decade. Along with all the travelling and writing and voluminous personal correspondence she also helped, in the early days after his death, to get William's books ready for publication. To make ends meet, she often rented out her house in Colville Road while she and the girls were abroad or spending summer weeks in Hindhead with the Pollocks or at Land's End with the Stephens.

The turning-point came six years after William's death when her reputation was made by the success of her novel, *Mrs Keith's Crime,* which is discussed later in the book. This success established her in the literary world and for many years thereafter her books and short stories sold well.

Marie Belloc Lowndes, Hilaire Belloc's sister and a prolific novelist herself, was another of Lucy's close and long-standing friends. She was unstinting in her praise of Lucy, and wrote that she deserved to be better remembered.[13] She makes the observation that Lucy's charming ways and serene demeanour concealed a 'deep-felt awareness of human unhappiness and a great compassion for it'. Mrs Belloc Lowndes, herself a popular hostess, made many friends through her visits to Lucy's modest home. It was there that she first met the young and unknown Bernard Shaw. She was at the first night of Lucy's play, *The Likeness of the Night,* and wrote showing her pleasure at its success. Over the years, she sent inscribed copies of her books to Lucy as they came out. Lucy's enthusiasm for the quality of Marie's storytelling rings true in the letters she wrote to her. They are not bland: she manages to present critical comments as signs of how interested she was in Marie's work.[14] With women friends, as with men, Lucy had a sensitive touch. The sensitive touch was totally absent in the letters Bernard Shaw wrote to Lucy. They had a long friendship. In 1928 she had sent him her play *A Woman Alone.* He replied with a friendly bombshell – 'It really is a horrible play!' He goes on to take the theme to pieces and ends:

> Why is it that all women who write, whether they are reactionary like you or progressive like — and — and —, all delight in leaving their readers thoroughly discouraged and miserable in the end?

He also burst off the page the next week in his reply to her rejoinder:

> I tell you *you* are the dupe of your own experience. Of course W.K.C. was cleverer than you. But he was cleverer than ME – cleverer even than Einstein . . . . So don't argue; but write another play that will not have for its proper title 'Back to the Eighteen-sixties!'[15]

But this was the year before Lucy died and she could no more escape from the limitations of her earlier themes than she could from the limitations of her own age. Shaw had always been interested in Lucy's work and once in 1893 he actually attributed the controversial and anonymously published play, *Alan's Wife,* to Lucy. He declared 'Whoever wrote that play it is of the Kingdom of Clifford from beginning to end – superstitious atheism – sensational anti-Goddity all over.'[16] His first choice guess was Mrs Clifford and his second choice Elizabeth Robins. Thirty years later it was claimed as the joint work of Florence Bell and Elizabeth Robins.

Lucy gave a picture of her day-to-day working life in an interview in 1899.[17] She was described as an intensely sympathetic and womanly personality who worked at her writing for long hours in the mornings and evenings, keeping the afternoons for tea and walking and 'desultory reading – so long as it is not indecent or improper'. The room she worked in was at the bottom of the house and her bedroom at the top. Lucy gave an amusing account about the fears that came upon her when she was writing late at night. She regularly fancied that she heard burglars 'about three quarters of an hour after the rest of the house have gone to bed', and was afraid to go upstairs to her bedroom. She had confided her fears to Thomas Huxley who replied that he too, eminent scientist and FRS though he was, suffered in the same way. He told Lucy that he 'not only heard burglars but actually *saw* them looking through the crack of the door!'[18] Thomas Huxley was twenty years older than Lucy and always signed himself in his letters to her as *Pater.* His letters to both William and Lucy are openly affectionate. He took Lucy under his wing when, soon after William's death, she had financial problems. 'Leave it all to be turned over in the mind of that cold-blooded, worldly, cynical old fellow who signs himself your affectionate Pater' he wrote; and Lucy must have been glad to do just that. Later, in 1895, when Lucy was pressing him to get a friend of hers elected to the Athenaeum, he sagely advised her not to meddle, and defended his advice in this way:

> Men, my dear Lucy are very queer animals, a mixture of horse-nervousness, ass-stubbornness, and camel-malice with an angel bobbing around unexpectedly like the apple in the possett, and when they can do exactly as they please, they are very hard to guide. How many men's chances have I seen smashed by their advocates! [19]

He was President of the Royal Society from 1881 to 1885 and was a tremendously popular public figure. Edward Clodd, the congenial philanthropist and friend of many of the most distinguised among the later Victorians, wrote that it was worth being born just to have known Huxley.[20] Clodd was much loved by his friends. For his sixtieth birthday in 1900 he was presented with a table specially made for the occasion. It was carved with the names of thirty of his closest friends. Lucy's name is there along with Thomas Hardy, Edmund Gosse, George Gissing, Frederick Macmillan, Holman Hunt, George Meredith, Herbert Spencer and other notables. Huxley's fiery early days, when he was known as Darwin's bulldog, and when his sparklingly provocative public lectures attracted the crowds, were finished. Nevertheless his regular attendance at Lucy's Sunday afternoon gatherings must still have been a magnet for other intellectuals and academics. Much earlier, in 1864, he had founded the powerful and élitist, nine-member *X Club* which flourished for twenty-nine years and was a major influence in the growing Scientific Establishment. Lucy's close relationship and loyalty to him was constant to the memory of the almost paternal care he had displayed for her husband, and, till the end of his own life in 1895, Huxley kept a practical, watchful and most affectionate eye on Lucy.

Notes

1. C. Morgan, *House of Macmillan*, 1843-1943, Macmillan 1943, p.148.
2. ALS (one of fourteen), Valehouse Collection.
3. ibid.
4. Eiluned Lewis (ed.), *Charles Morgan: A Memoir*, Macmillan, 1967, p.19.
5. As note 2.
6. ibid.

7. Unidentified Obituary by 'A Friend' 1929.
8. ALS, Valehouse Collection.
9. Hugh Walpole, *The Apple Trees: Four Reminiscences*, Golden Cockerel Press, 1932, p. 45-46.
10. Theodora Bosanquet, *Personal Diary,* 1914-1918, Unpub. Harvard University Library (96-1918).
11. ALS (one of five), Valehouse Collection.
12. Sir John Pollock, *Time's Chariot*, John Murray, 1950, p.80.
13. Mrs Marie Belloc Lowndes,*The Merry Wives of Westminster*, Macmillan, 1946, p. 62.
14. Fifty ALSs from Lucy Clifford including nine to Mrs Belloc Lowndes are held at the Harry Ransom Humanities Research Center, Austin, Texas.
15. ALS (one of two), Valehouse Collection.
16. Dan H. Laurence (ed.) *Bernard Shaw: Collected Letters*. Reinhardt, 1965. Letter to Elizabeth Robins dated 28 April, 1893, p. 393.
17. Mary Angela Dickens, *A Chat with Mrs W.K. Clifford*, Windsor Magazine 1999. Vol 9, p. 483-485.
18. ALS (one of six), Valehouse Collection.
19. ibid.
20. Edward Clodd, *Memories*, Chapman and Hall, 1916, p.40.

Friendship with William and Henry James

Before Lucy met Henry James she had known his brother, William, who held a highly respected position in academic and intellectual life in America. At the time when Henry was travelling in Europe and establishing himself as a writer, he was regarded as something of a lightweight by comparison with his brother. From his letters it is clear that Henry often sought, but did not always receive, his older brother's approval for major decisions in the way he ran his life. William sometimes tended to take the 'elder and better' role in his dealings with Henry, but they did agree about Lucy Clifford: they both responded warmly to the particular attraction of her personality.

Although William James and Lucy met on only a few occasions early in the 1880s, his letters to her show that he enjoyed her company. Lucy would have been drawn to anyone who was an admirer of her beloved husband, and William Clifford's philosophical writing had certainly made a great impression on William James. William James makes much of Clifford and his views. He writes: 'What we enjoy most in a Huxley or a Clifford is not the professor with his learning, but the human personality ready to go in for what it feels to be right, in spite of all appearances.'[1] William James delighted in Clifford's lack of respect for religious conformity. He called him 'that delicious *enfant terrible*'[2] although he could not go all the way with Clifford's uncompromising and, to many, uncomfortable assertion that:

> Belief is desecrated when given to unproved and unquestioned statements for the solace and private pleasure of the believer.... Our duty is to guard ourselves from such beliefs as from a pestilence which may shortly master our own body and then spread to the rest of the town.... It is wrong always, everywhere and for every one, to believe anything on insufficient evidence.[3]

In his essay on *Great Men and their Environment,* William James, discussing his proposition that a community is a progressive and a living thing, writes:

> I can do no better than quote Professor Clifford: It is the peculiarity of living things not merely that they change under the influence of surrounding circumstances, but that any change which takes place is not lost but retained, and, as it were, built into the organism to serve the foundation for future action. No one can tell by examining a piece of gold how often it has been melted and cooled in geologic ages, or even

in the last year by the hand of man. Anyone who cuts down an oak tree can tell by the rings in its trunk how many times winter has frozen it into widowhood, and how many times summer has warmed it into life. A living being must always contain within itself the history, not merely of its own existence, but of all its ancestors.[4]

William James and his wife were fond of Lucy, and certainly William visited her at Colville Road. In 1883, four years after her husband's death, Lucy had sent William James one of her earliest short stories which had been published anonymously. She begged him not to reveal her as the author: she sought no recognition; in fact, she feared that her late husband's friends would scorn her for the sentiments portrayed in the story. There seemed little chance that an unknown young woman's writings would impress this distinguished psychologist and pragmatic philosopher from Harvard. Yet Lucy Clifford's story did capture his interest and stir his emotions. She had written a heartrending tale of the spirit of a recently dead young woman seeking to make some contact with her husband and child. Her anguished fear is that her husband's second wife might not love the child. Lucy, as the spirit of the dead woman, writes:

> It is only in dreams that the dead have power over the living, for theirs is the land of which the living see only fitful gleams in their sleep – a land where, to the living, all seems and nothing is, and nothing earthly has an abiding place . . . . I stole in the darkness through the quiet house, and found the room where the child lay sleeping . . . . I saw its face – its soft hair and closed eyes, and heard the sweet sound of breathing that came through its parted lips – and I longed for human life again, and would have given my soul up thankfully to have had my flesh and blood back for one single instant, to have held that little one in my arms . . . . Did she love it? If I might know that, I would be content, but nothing had a hold on me – there was nothing I could reach again in the human world . . . . I passed out into the night – on and on – farther and farther away – seeking the infinite and finding it never . . . .[5]

William James wrote back:

> Your story is a very wonderful and beautiful thing, and has made a singular impression on me . . . . I doubt if ghost-life has ever been as intensely imagined. I shall write to my wife to get it but keep your anonymity to everyone else since you wish it . . . . When I return from Paris I hope I shall have the pleasure of finding you at home again.[6]

The letter was a splendid encouragement for a young and relatively unknown writer. William James really did write to his wife about the story, and nearly forty years later his daughter Margaret, going through her mother's papers after her death, came across the letter and copied part of it for Lucy, also telling her that he 'greatly and unreservedly liked Mrs Clifford':

> Mrs Clifford sent me a story of hers called 'Lost' in MacMillan's (sic) magazine for May 1881. Don't fail to get it from the library ... it seems to me the most imaginative Ghost and immortality story I ever read. Obviously she cares much about immortality, but thinks it her duty to care nothing for it. Don't tell anyone she wrote it; she seems in deadly fear lest Leslie Stephen should find it – but you will understand when you learn that he and his second wife are her intimate friends and when you read the thing.[7]

Lost remains one of the most evocative short stories Lucy wrote, and, remembering that it was written soon after the death of her husband, the desperate situation described in it has a terrible poignancy. Two years later William James, writing to Lucy from his Appian Way home in Cambridge, Massachusetts, includes two sentences which bring perhaps a hint of the affinity between them. He mentions that he has not forgotten their talks together and goes on to say that the metaphysician in him does not understand why there seems to be something 'doubly friendly' in tokens of recognition that come, as he puts it, from 'your dark, old world beyond the sea. I don't know why it should be, for a human heart is a human heart, wherever found – but it *is.*'[8] He had been deeply affected by her story, but one senses that Lucy herself had found a corner in his heart.

Lucy first wrote to invite William James's younger brother Henry to tea in March 1892.[9] But their paths were crossing much earlier, and he may well have met William Clifford in the months before his illness became severe. From late in 1876, when he was lodging in Bolton Street, enjoying the 'Season' in London, and youthfully bustling round Europe, Henry James was often invited to Thomas Huxley's home in St John's Wood. There he met again Leslie Stephen and Julia and, through them, the Thackerays. He knew Frederic Harrison and Robert Browning and Lord Leighton – all of them friends of the Cliffords. Henry was also often at the Pollocks, and took his brother William there to introduce him into London society. When he did meet Lucy he thought, with everyone else, that she was many years younger than her husband. In fact she was only one year younger, and thus only three years younger than Henry James himself. Leon Edel states that James and Lucy first met in 1880 and that may well be so, but he offers no dated corroborative evidence:

> Lucy Clifford ... had been in her youth a golden-haired, red-cheeked art student, who sketched antique statues in the British Museum .... She was still wearing mourning when James met her in 1880 .... By the 1890s ... she had become a hearty, mothering, energetic, enveloping woman, direct in her conversation and formidable in her ability to get things done .... Unlike the ladies James had cultivated in his younger years, she did not possess a sharp tongue, or an underlying cruelty that he had accepted and even found attractive in them. Lucy Clifford was genial. She spoke well, she was eminently expressive and generous ... . She had loyal friends among London's most distinguished men .... James would speak of 'that admirable Lucy Clifford' as 'a character,

a nature, a soul of generosity and devotion'. He liked nothing more than to sip a liqueur by her fireside . . . or to take tea with her and share London literary gossip.

Whatever the details of their first meetings were, it is clear that Lucy got on well with both the brothers, but it was with Henry that friendship grew in strength and intimacy until, at the time of his death in March 1918, she was one of his closest London friends.

The letters he wrote to Lucy reveal how much James treasured the closeness of their friendship. Sadly, few of Lucy's letters to him remain: they probably went up in smoke in his regular burnings of personal correspondence in the fireplace and garden at Lamb House. In most of his letters he addressed her as 'Dearest Lucy C' or 'Dearest Lucy' but used 'Good and Only Aunt'; 'Dearest Creature'; and 'Beloved Girl' too. Almost every letter is closed differently – 'your all faithful and fond'; 'yours re-boundingly'; 'evvy your Nevvy'; 'your ever so affectionate'; 'your doting old Nevvy'; 'constantly yours'; 'your most adhesive and devoted'; 'your ever clinging'; 'your constant old Nevvy'; 'tender and true'; 'all increasingly'; 'seedy and stumbling but not quite crumbling' Henry James. Just one letter ended up with a frivolous 'ever your Lambkin'! The '*Aunt*' and '*Aunt Lucy*' and '*Nevvy*' epithets came about when, in 1892, Lucy had written a book, considered to be one of her best, with the title *Aunt Anne*. He started the teasing appellation and Lucy reciprocated.

Henry James enjoyed female company and the comfortable world of 'at homes' and luncheons, theatre visits and leisurely European travel suited him down to the ground. Edith Wharton, the wealthy American writer, invited him to travel both in Europe and the States. He loved the luxury of her homes and thoroughly enjoyed her invitations for outings in her expensive motor cars. He was never bored with Fanny Prothero, his lively, good-neighbourly friend and advisor, whose husband was editor of *The Quarterly Review*. Elizabeth Robins, the colourful actress and novelist, was a close friend. She saved the letters he wrote to her and used them later in her book *Theatre and Friendship*. Jessie Allen was his 'lady bountiful' to whom he wrote more than two hundred letters over the seventeen years of their friendship. The novelist Rhoda Broughton was a close friend too, and of course there was his mysterious relationship with Constance Fennimore Woolson which ended with her violent suicide in Venice. But Lucy Clifford always held a very special place in his affections. He wrote of her 'great personal sweetness and frankness and niceness, and her heroic and arduous life, her admirable courage and labour and gaiety and independence through it all.'[11] He went often to her Sunday teatime Salon, and, along with her young literary protégés and older established figures, she would briskly organise him, sort out his domestic problems and treat him like the other, much younger, writers she loved to encourage and counsel. He sought her company and was comfortable enough in it to accompany her to her dressmaker, wait for her, and take her out to tea. He would lunch with her at her club – the Ladies' Athenaeum in Dover Street. He would take her to the West End Cinema and walk back with her to her home near Paddington Station.[12]

In his letters, Henry James often adopted a very affectionate style of address to his friends. However, his letters to Lucy reveal the fact that he was, above all, completely at ease with her. She did at times irritate him with her 'managing' of his affairs; but his deep-seated sense of politeness probably led him to disguise any irritation she caused him. However, in one letter when he was depressed about the progress of the war, – especially the overrunning of Belgium – he was driven to speak out plainly to her from his home in Rye:

> To this end, very kindly, don't send me on newspapers – I very particularly beseech you; it seems to suggest that you imagine us living in privation of, or indifference to, them: which is somehow such a sorry image. We are drenched in them and live up to our neck in them: all the London ones by 8am and every scrap of the evening ones by about 6.40.[13]

Lucy was always short of money and time, but must nevertheless have taken it upon herself to bombard him with newspapers. The incident is typical of Lucy in her managing mode. Much earlier, when his play *Guy Domville* was staged at the St James's Theatre in London in 1895, Henry James had endured a humiliating experience: he had appeared on stage at the end of the play only to be booed and catcalled off it. He was deeply shocked. The whole experience of being made a public object of scorn, even though it was seen later to be partly a contrived confrontation, cut deep into his self-esteem. Sitting in the front stalls was a glittering array of famous-to-be names: the young H.G. Wells, Arnold Bennett and Bernard Shaw were cutting their literary teeth as theatre critics. Lucy was in the audience with a great crowd of Henry's friends. Their applause and appreciation was overpowered by the vociferous cat-calls from a group who seemed determined to disrupt the proceedings by audible manifestations of their contempt for certain parts of the play. In the following weeks he needed the comfort of his friends and he sought special sympathy in Lucy's comfortable upstairs sitting room. He spoke often of times spent there when Lucy would settle him by her fireside, open her cupboard and draw out the Benedictine bottle, chat with him and smooth his ruffled feelings. At this time of his keen disappointment about the lack of success for his theatrical ambitions, Lucy again took up her advisory role. In 1896 she had urged him to write more popular, more readable stories. She approached her friend Clement Shorter, who was then editor of the *Illustrated London News,* and persuaded him that Henry James had up his sleeve the idea for a story that could be published in instalments and which would be attractive to his readers. Shorter duly approached Henry James who replied to him in business-like terms. As far as financial arrangements were concerned, he seemed to adopt a crispness and directness of style but with definite Jamesian touches:

> I should be very glad to write you a story energetically designed to meet your requirements of a 'love story' – and to let you have it at the time and of the dimension that you mention – but the sum you name is less than I am in the habit of receiving: it is, in fact, rating the instalments, individually, at my usual fee, a great deal less. I should however like to

capture the public of the Illustrated News, and should be glad to surrender to you the serial rights of the work in question for £300, if you see your way to meeting me on that figure. The particular story I conceive Mrs Clifford to have spoken to you of is – must be – a plan I narrated to her more than a year ago and have carried for longer than that in my head: an idea that would, as I remember telling her, lend itself about equally to a play – 'of incident' or to a novel – of the same. Two girls are indeed in the forefront of it. If you accept my amendment to your terms I should be able presently to fall to work on it. Believe me yours very truly Henry James.

Henry converted to a novel a story which he had originally thought of as a play, and it was indeed published in the *Illustrated London News* later in *1896*. It appeared in thirteen instalments from 4 July to 26 September under the title *The Other House.* Lucy would have been well pleased with her intervention on behalf of her famous friend. Henry James later re-wrote the story as a play but it was never produced. This experience with Shorter seems to have been a mixed blessing to the novelist. He had no great respect for the *Illustrated News* and in a sense he tailored the melodramatic story with, for him, an unusual element of violence, for the journal and its readers. The experiment was not repeated.

In several of Lucy's novels the influence of Henry James's writing can be detected. She took up some of his themes to provide elements for some of hers and the fact that he must have been aware of this but unconcerned about it shows how mutually tolerant their relationship was. It is suggested that Lucy portrayed James as the character Henry Langton in her 1901 novel *A Woman Alone.*[15] Likewise, some critics have seen a portrait of Lucy Clifford in James's short story *Greville Fane.*[16]

All of James's women friends were wealthier and probably more sophisticated than Lucy Clifford. She was busily occupied with her journalistic work and her gossip column and her novels, and at times she invited friends to an informal lunch announcing in advance that it might consist of 'cold turkey bones and not much else'. Nevertheless, Henry James was always ready to spend time with her at her modest Chilworth Street home where he was sure to be warmly received. He would sometimes tell her in advance – no doubt to spare her purse – that he needed only 'a very small cutlet'.

When Henry James first lived in London he took rented rooms in Bolton Street off Piccadilly. From 1886 he had a comfortable and well-placed flat at 34, De Vere Gardens, Kensington and later moved to Carlyle Mansions in Chelsea; but, in 1896, he fell in love with a house in the country. For holidays he had sometimes visited the Romney Marsh area of East Sussex and discovered the pearl in its mists, Rye. Even today, when tourism has busied the streets and changed the atmosphere that it would have held at the turn-of-the-century, it is still easy to imagine the delight that the quaint almost mystical mound of red-roofed houses, nestling along steeply winding cobbled streets, would present to a visitor from London. The house he fell in love with was Lamb House which he had noticed and admired when he had rented another property nearby. In 1897 James managed to purchase, for £70 a year, a 21-year leasehold on this house of his dreams. He had never thought it would come on the market but, by chance, he was the first to know that it would become available and he immediately decided it must be his. The house is unusually placed at the inside angle of the corner of two streets, and from either approach the red brick building is most attractive. Earlier pictures of it, showing the Garden Room before it was destroyed in an air-raid in 1940, allow us to understand why the author coveted it. When he had finally acquired it he gave his friends a blow-by-blow account of the renovation and furnishing of it. He felt that he would happily spend the winters there for the rest of his life. In 1899 he was offered the freehold. He agonised over the price – £2,000 – and he wrote to his brother William, enthusing and hoping to win his approval for the purchase.[17] But William poured cold water on his plan and cast doubt on the wisdom of the investment, and Henry became once more the younger brother on the defensive. He remained determined however, and Lamb House became his.

An interesting aspect of becoming the owner of an out-of-town property was that a deeper layer of his complicated personality was exposed to the friends who were invited to stay. To see him away from the London drawing-room setting, and on his own territory, was to cut a little deeper through the smoke-screen of stylised behaviour he generally assumed. To be met at the station, to walk in the garden, to sleep in the guest bedroom, to join him at the breakfast table, to share his house, to walk the streets of Rye with him, all these interactions gave his friends more insights into the real Henry James than the structured London life had ever allowed. He created and sustained a new role for himself, and, from all accounts, he settled into his new territory with great style. In Rye he became the country gentleman. Visitors like Edmund Gosse, one of his closest friends, were amused to report the proprietorial manner in which he would swoop upon his friends as they emerged from the train at Rye station, take possession of them and their luggage and escort them ceremonially through the streets of *his* town to *his* house.[18] He presented a slightly eccentric and imposing figure to the local townsfolk. He was fastidious about his dress and something of a dandy in his choice of garments and would certainly not pass unnoticed in those attractively narrow winding streets. He, and the colourful visitors he attracted, became a talking point in the town. Conrad, Chesterton, Kipling, H.G. Wells, Max Beerbohm, Violet Hunt, Edith Wharton, Ford Madox Ford, Hugh Walpole, Rupert Brooke and Percy Lubbock were among those who could be spotted about the ancient buildings and quaint corners of the town. Besides ambling through the streets with his habitual walking stick, the author could be seen on his bicycle in his distinctive knickerbocker suit, cycling to visit Stephen Crane and his Cora at nearby Brede, or simply taking exercise on the flatter, traffic-free roads of Romney Marsh. Crane, the young American novelist, best known for his book *The Red Badge of Courage*, was among the band of writers who lived or had homes in East Sussex during some of the twenty years that James owned Lamb House. However, he found himself missing the vitality of his London friendships and, after the winter of 1899 when he had longed for London and felt homesick for it, he wrote in May to Lucy:

> But can't you, couldn't you, dearest L.C., come down and see me some day next month, some pretty, summery, possible day as you did two years ago, and tell me everything that's going and let me drive you to

Winchelsea, after feeding and tea-ing and blessing you? Think of it, plot for it – and let me hear of it. It seems to me as if otherwise I mightn't see you for many a month . . . it would be infinitely nicer to be sitting by your fire and tasting your charity – and your Benedictine.[19]

It was an irresistible invitation to repeat a delightful earlier visit when he and Lucy had shared 'the inside' of a day and Henry had feasted on Lucy's graphic descriptions of what he was missing in London. By a strange twist of fate, the ownership of Lamb House played a part in his decision to seek, in 1915, naturalisation as a British subject. He certainly wanted to show his solidarity with Britain at that time of the war, but, as Theodora Bosanquet reported, 'He might not have troubled to take the step just then if he hadn't heard that he couldn't go back to Rye in the usual easy way. As an American alien he would have to be registered and placed under the observation of the police.'[20]

Lucy Clifford had presented inscribed copies of her books to Henry James, but they are not on the shelves at Lamb House now. Much of the library was sold when the house was left to the National Trust in 1950. Some of her books may have been lost among the 200 or so destroyed in the bombing of the Garden Room. Neither does Lucy feature in any of the photographs which remain there. However, the fact that Lucy's husband is today famous for his mathematical discovery, known as *Clifford Algebra*, now furnishes this ditty, which hangs framed in one of the rooms at Lamb House, with a frivolous appeal. The origin of the verse and how it came into James's possession is not clear:

> In heaven there'll be no algebra
> No learning dates or names.
> But only Angels playing harps
> And reading Henry James.

When he was in London, Henry James often visited Lucy. He enjoyed walking, and in the early days from De Vere Gardens, he could walk across to the north side of Kensington Gardens to reach Colville Road where she lived till 1899.[21] In 1905, during the hectic time when Lucy was preparing for the wedding of her adored elder daughter Ethel to Fisher Wentworth Dilke, Henry James was in America. He generously lent her Lamb House so that she could relax for a time with the engaged couple. During her stay Lucy wrote to him. One of Lucy's most attractive qualities lay in the fact that she seemed always to include in her letters exactly what her reader wanted to hear. This previously unpublished letter gives us a glimpse into the domestic round at Lamb House. Henry would have read it greedily at Chocorura in New Hampshire and relished the reminders of his English home:

May 4th 1905.
Beloved H.J. –
   We are happy. I begin at once with that. Lamb House is [more] than adorable. Everything is in order and going beautifully. Mrs Paddington is a treasure, and all the servants are excellent. 'The boy' [Burgess Noakes

who was taken on as house boy in 1901 and stayed devotedly serving H.J. till his death.] is really a joy. His butler-like airs quite subdue us. Fisher said he behaved beautifully, brushed his clothes, laid them out, and behaved like a full fledged valet. Max [H.J's dachshund] is quite well, we grieve to say that we consider him somewhat an impostor, for one day he appears to be unable to live without us, and cries outside the door if he is not at once admitted – he goes to sleep and snores loudly, on my lap. The next, he forgets all about us – the next suddenly he tears upstairs and salutes all round as if sorely against his will he had been kept away from us. Mrs Paddington, I advise you never to part from while you live. She is a miracle of frugality and forethought. I declare that living here with Fisher added, and at least two more in the kitchen than we have at home, the books come to less than ours do in town. Of course the fact that there are no droppers-in to luncheon and dinner may make some difference but not enough to detract from her accomplishments. We sit mostly in the upstairs study. When the young people wanted to be alone we had a fire in the drawing room. But Fisher went this morning . . . . We shall weep to go the following Saturday. But it will leave us just one month to do everything for the wedding. Turkey [Lucy's younger daughter] comes home on the 10th. I shall have to get not only Ethel's bridal array and going away dress (thank Heaven all the other things are bought or being made) but a wedding garment for Turkey, for myself, all sorts of things to do. And a ground-floor lodging outside has to be found for Turkey who says she is to be considered well but she can't get up and downstairs yet. There is no downstairs room available at Chilworth St. I must put her opposite where Lisa stayed. Your knocker is a constant joy to us. The Fishermaid and man could let themselves in and out so easily.

She mentions the splendid, ornamental brass front-door knocker since it is unusual in that it can also be turned to open the door. Lucy met several of

> Henry's Rye friends and she does not neglect to flatter him as she continues: We all talked of you, boasted of the intimacy with you, of how much we loved you, of your works and your triumphs (of which all literary England boasts and takes as a national compliment) and they told stories of your life in Rye . . . . A newspaper was brought out with a portrait of you, (benevolent looking but not flattering) . . . and extracts from Julian Hawthorne's interview. I delight at your description of California. I am sure that you have seen it truly. Your book will be splendid. It will make your fortune!! . . . I feel as if this letter were all about ourselves – but in our present setting you will forgive it.– Much love to you from yr. affectionate and happy and [smudged word here] friend Lucy Clifford.[22]

Henry would have been delighted to receive such a letter. He adored his dog Max, 'the best and gentlest and most reasonable and well-mannered as well as the most beautiful small animal of his kind'. He had been a replacement for the terrier, Nick, who died in 1902, and who had become the fourth dog to be immortalised by the inscribed stone slabs that can still be seen on the wall in a shady corner of the garden at Lamb House.

In 1907 Vanessa Stephen married Clive Bell. The Stephen family were still grieving the deaths of Julia, Stella, Leslie and Thoby. Henry James expressed his devastating opinion of Clive Bell in his 17 February letter to Lucy:

> I went to see Vanessa Stephen on the eve of her marriage (at the Registrar's) to the quite dreadful-looking little stoop-shouldered, long-haired, third-rate Clive Bell – described as an 'intimate friend' of poor, dear, clear, tall, shy, superior Thoby – even as a little sore-eyed poodle might be an intimate friend of a big mild mastiff. However I suppose she knows what she is about, and seemed very happy and eager and almost boisterously in love (in that house of all the Deaths, ah me!) and I took her an old silver box ('for hairpins'), and she spoke of having got a 'beautiful Florentine teaset' from you.[23]

This was just part of a long letter containing other indiscretions in which, in an earlier part, he had enjoined Lucy to 'repeat me, quote me, betray me not – and burn my letter with fire or candle (if you have *either*! Otherwise wade out into the sea with it and soak the ink out of it.)' Lucy did not betray the trust he had in her, but she did save the letter amongst the collection that she passed to Percy Lubbock. He told Lucy that he would not use all the letters that she had given him since James was 'unprintable when he was at his best' and that perhaps in a hundred years they would eventually be published![24] This particular one was included in Leon Edel's selected collection in 1974.

On 4 September, 1910, Henry James wrote to her from Chocorua. His brother William had died just days before on 26 August. Lucy had cabled her sympathy and must somehow, in those few typescript words on an international cable-slip, have so transmitted the warmth of her feelings that Henry replied quickly and gratefully for her 'so immediate and tender and *participating* cable'. The letter of 4 September is a particularly interesting one. It demonstrates the empathy between Henry and Lucy and shows that, consumed by anguish as he was, he recognised that Lucy had suffered too and associated his grief with hers. Here is part of that letter:

> Sept 4 1910, Chocorua, NH, USA.
> I was much moved in my deep anguish by your so immediate and tender and participating cable. We sit here stricken and in darkness, and the sense of these last miserable weeks is as a black nightmare to me. My beloved Brother died, after an amount of suffering it had been anguish to behold – helplessly – just a week after his arrival here; having got swiftly and deplorably worse during our voyage – which, however, had been extraordinarily fair and rapid. I can't speak of what his extinction means for me – it has made me, for the time, feel myself old and ended. He was not only my cherished and admired Brother –

but my best friend in the world my great elder and better, & authority as he had been from far back in the distant years of childhood and he was at the very climax of his beautiful genius, his high intellectual life. Who should know better than you, however, what this particular sort of pang and pain means – you who, a long time ago went through it, and worse, as W.K.C. was taken so much earlier in a great career, and who have yet faced life so greatly yourself ever since? Think of me none the less tenderly – for the difference intimately made to me is unspeakable . . . I dare say you will have written me a word, eight days ago, and given me of your news, a little, whatever it is. I shall rejoice, dearest Lucy, in every echo of you and am more than ever your faithfully fond old Henry James.[25]

Lucy followed up her cable with two letters which Henry felt to be 'quite divine'. However, he delayed his reply so long that he felt he needed, on 29 October, to write that:

My silence has been atrocious, since the receipt of two quite divine letters from you, but the most particular blessing of you is that with you one needn't explain nor elaborate nor take up the burden of dire demonstration, because you understand and you feel, you allow, and you know, and above all you love (your old entangled and afflicted H.J.)

He continued with this moving tribute:

Your letters meanwhile, dearest Lucy, were admirable and exquisite, in their rare beauty of your knowing, for the appreciation of such a loss and such a wound, immensely what you were talking about. Every word went to my heart, and it was as if you sat by me and held my hand and let me wail, and wailed yourself, so gently and intelligently, with me. The extinction of such a presence in my life as my great and radiant (even in suffering and sorrow) brother's, means a hundred things that I can't begin to say . . . . We will talk of all these things by your endlessly friendly fire in due time again (oh how I gnash my teeth with homesickness at that dear little Chilworth Street vision of old lamplit gossiping hours!) and we will pull together meanwhile as intimately and unitedly as possible even thus across the separating sea.

He ends:

I do yearn after you, and I needn't tell you how any rough sketch of your late history will gladden my sight. I wrote a day or two ago to Hugh Walpole and besought him to go and see you and make me some sign of you . . . . I think of you as always heroic – but I hope that no particular extra need for it has lately salted your cup. Is Margaret [Lucy's second daughter] on better ground again? God grant it! But such things as I wish to talk about – I mean that we might! But with patience the hour will strike – like silver striking silver. Till then, I am so far -offishly and so affectionately yours, Henry James.²⁶

That he longs for Lucy's sympathetic company is made quite clear in these lines. His image of 'silver striking silver' was echoed in Lucy's own gravestone epitaph about herself and her husband.

In February of 1912, Lucy visited the United States. The trip had been arranged by Scribner and she carried with her various introductions, including one from Henry James, who wrote to a friend 'hold out your hand to her for my sweet self – almost as if I were in love with her and you wished to be sublime'. She was in New York when the famous Harper's New York Dinner of 1912 took place. This grand affair had been set up by the publishing company to honour William Dean Howells on his 75th birthday. The friendship between Henry James and William Dean Howells was remarkable. It lasted for the whole of their lives and is recorded in the many hundreds of letters which passed between them. As young men – James was just six years younger than Howells – they bounced their precocious literary ideas between each other as they walked, talked and dreamed together. Howells went on to succeed as a novelist, became editor of the *Atlantic* and later was known as the 'King of Critics'. And now, for this anniversary dinner, Henry had written an open letter to his old friend. He had intended that his eulogy should have been read out at the dinner but was afraid that it had not reached him. Lucy Clifford was invited as a guest of honour at the dinner. She wrote to Henry from New York telling him of the event. Henry knew that when she returned she would provide the sort of vivid, unrestrained descriptions of the evening that he loved to hear. He wrote to another boyhood friend, Thomas Sergeant Perry:

> An old friend of mine here, Lucy Clifford (W.K.'s widow) who was there and at the 'High Table', has sent me a catalogue of the guests which I hang over in the appallment of fascination – or the fascination of appallment; and which, as she has just returned and I am to see her tonight, she will fill out with hideous detail . . . the weird desolation of it.²⁷

Henry knew that the year before, Howells, together with Edith Wharton, had mounted an unsuccessful campaign to get the Nobel Prize awarded to him. The whole text of James's 'open' birthday letter was published in the April edition of the *North American Review*. It was a moving tribute to the man whom he had characterised as Lambert Strether in *The Ambassadors*. Thomas Sergeant Perry had felt that Harper's had somehow managed to vulgarise the event to the point of embarrassment. This had whetted James's appetite for gossip and Lucy's reporting of the evening would have been infinitely interesting to him. She had a good eye and ear for interesting details which had earned a mild criticism from Leslie Stephen, who found her style of writing 'a little too much immersed in the journalistic element'²⁸ for his taste. His daughter, Virginia Woolf, was more brutal, for she cruelly labelled Lucy 'A hack to her fingertips.'²⁹

Lucy responded to Henry James's moods, rushed to his assistance when he was ill, pushed him into accepting invitations, and, most importantly, saved special snips of gossip for him. They were good for each other and good to each other. One of his fears, when living at Rye and missing his evenings at her fireside, was that she would be 'letting off steam by which I shan't – or don't – profit. Do bottle a little up for me, and I will come and uncork the wine on the earliest possible day.' He did seem sometimes to yearn for Lucy and was able to straightforwardly write that it was 'horrible not to see you immediately'. They were secure with each other. Each of them at times needed to forgive the other when they had fallen short of the other's expectation, but there was no jealousy in their relationship.

That he trusted Lucy implicitly was evident from the fact that he allowed himself quite blisteringly critical thrusts at other theatrical and literary personalities in his letters to her. If she betrayed him he was lost. She sent him letters that she had received from friends and which she thought would interest him. With each other they were able to indulge themselves greedily, perhaps even gluttonously, in gossip. They both proved their love for the other by each willingly placing in the other's hand a weapon by which damage could be inflicted – the confidences of others. As far as is known there was never a betrayal. For example, in 1900, Lucy had written him a letter full of vivid details involving marriage, widowhood, pensions and Christina Rogerson of the infamous Dilke divorce scandal of 1885-6. Christina had been a long-standing friend of Henry James and he wrote back on 4 January to declare that Lucy's 'vivid and copious letters' . . . felt to him 'as if every written word from you is a drop of your blood, and I value it with a tragic appreciation – deprecating even while I devour'.[30] With Lucy he was even secure enough to go on with this most revealing perception of his own vicarious interest: 'I feel with you, like one of the augurs behind the altar – winking at the other. But it's I who do the winking – you are perfectly proper.' This relishing of exchanged scandalous gossip could seem shameful, except for the fact that here we are seeing Henry James the voyeur and Lucy Clifford the raconteuse, safely at play and harming no-one. Both of them were intensely interested in the London scene. When Henry was at Rye, he was starved of news and Lucy simply fed him it. He sent her flowers from his garden; she sent him news from London. On one occasion he enclosed with his message to her a reproachful letter which he had written to a mutual acquaintance and which he wanted her to address for him. He purposely left the enclosed letter open for her to read and enjoy before posting it. As an afterthought, fearful of what would happen if she inadvertently put the wrong letter in the envelope she was to post on, he begged her to BURN *his* letter to her immediately after reading it. In many, many letters, Henry James writes thanking Lucy for the immediacy of her warm, generous response to him. He too responded quickly when he felt Lucy was in need of sympathy and support. He wrote her this letter when he had heard that one of her daughters was threatened with serious illness:

> My dear, dear, dear Lucy! Never have I loved you so much! You have the grandest, noblest, clearest, strain of courage that ever was – never fear that it will fail you. Here your natural fears try to assault you – but

you will keep even them (in which time will immensely help you) at bay! And we live through everything and everything somehow ministers to life. Ah, how close to you I should like you to feel that I stand! And ah for tomorrow's talk! I shall really come almost (quite) at 8! All your H.J.[31]

No conventional commiseration here: simply participation with her and confidence in her. Through his letters to Lucy Clifford we see a warmth in Henry James perhaps not seen elsewhere.

Lucy was supportive to her friends but her letters to them were not bland. She was not afraid to say what she thought, and, in particular, one of her letters had a most important consequence. In 1913, when Henry James's seventieth birthday was approaching, a group of ten close friends, under the leadership of Edmund Gosse, Percy Lubbock and Lucy, decided to circulate his friends for subscriptions to a birthday fund. They wanted to commission a portrait of James from John Singer Sargent and to purchase a substantial commemorative gift. The plan was meant to be kept secret from Henry James but, when he got to know of it through Jessie Allen, he wrote to Percy Lubbock asking him to put a stop to it. Lucy intervened on behalf of the committee and Theodora Bosanquet recorded in her diary that Henry James:

> received fifteen close-written pages from Mrs Clifford, attacking him for his action and accusing him of 'coldly, callously and ungraciously rejecting the gift of his friends' and 'generally making an appalling fuss!' He, poor dear, was so overcome . . . that he could speak of nothing else for the first part of the morning.[32]

Lucy's letter won the day and ensured that the birthday gift they had chosen, a silver-gilt 'golden bowl' was presented and that the splendid portrait, regarded as one of Sargent's finest, now hanging in the National Portrait Gallery, was painted. Lucy wrote of the actual birthday celebrations to Charles Scribner. She described the scene at Carlyle Mansions where she found James: 'bewildered – his staircase a sort of highway for messengers and telegraph boys carrying messages, wonderful flowers arriving, the telephone bell going like mad . . . .' Lucy Clifford's role on the day itself was to make sure James survived the unusual hustle and bustle at his home. Afraid that the 'delightful lunacy' of it all would prove too much for him, she told the Scribners that she 'carried him off to lunch peacefully at my club and then returned him to the rejoicings'. He was perfectly delighted with the portrait and described it to Rhoda Broughton as:

> Sargent at his very best and poor old H.J. not at his worst; in short a living breathing likeness and a masterpiece of painting. I am really quite ashamed to admire it so much and so loudly, it's so much as if I were calling attention to my own fine points. I don't, alas, exhibit a 'point' in it, but am all large and luscious rotundity – by which you may see how true a thing it is.[33]

Sargent's portrait of Henry James, who described it as 'Sargent at his best and poor old H.J. not at his worst'. Lucy Clifford took a leading role in commissioning the seventieth-birthday commemorative gift. The portrait now hangs in the National Portrait Gallery in London.

All his earlier misgivings were forgotten and his letter of thanks to the two hundred and sixty-nine subscribers to his birthday fund reflected the 'boundless pleasure' he had derived from the celebration.[34]

Active loyalty to her friends was Lucy's strong point and Theodora Bosanquet notes how very attentive Lucy was during Henry James's last illness. She was popping in to Carlyle Mansions almost daily, kindly bringing roses for Theodora and making domestic and other decisions with Mrs James. Lucy had been one of the group who had tried to arrange for the Henry James's funeral to be held in Westminster Abbey, but the Dean of Westminster told her that this would only be possible at the request of Crown or Government. It was also typical of Lucy that, knowing how very disappointed Henry James had been at his lack of success in the theatre, she tried to redress the balance and to set up a dramatic triumph for him after his death. She asked Theodora Bosanquet to assist her in her idea of having *The Other House* put into playable order by Sir Arthur Wing Pinero and produced soon after the funeral.[35] Sadly, the idea came to nothing. Our memory of Lucy is clearly enhanced by her being linked with 'the Master' but she, by her friendship and influence, has enhanced also our memory of him.

Notes

1. William James, *The Will to Believe*, University Press, Cambridge USA, 1897, p. 93.
2. Ibid, p. 8.
3. William Clifford, *Lectures and Essays*, Vol. 2, *Ethics of Belief*, p 183-6.
4. ibid, vol.1, p. 82.
5. Mrs. W.K. Clifford, *Lost*, published anonymously in *Macmillan's Magazine*, May 1881.
6. ALS, Valehouse Collection.
7. ibid.
8. ALS, March 10 1885. Valehouse Collection.
9. M. Demoor and M. Chisholm, (eds.), *"Bravest of women and finest of friends": Henry James's letters to Lucy Clifford.* English Literary Studies, University of Victoria, 1999, items 77 and 78.
10. Leon Edel, *The Life of Henry James*, Rupert Hart-Davis, 1972, vol. 5, p. 106.
11. *The Pocket Diaries of Henry James* contained in *The Complete Notebooks of Henry James, 1910-1915* edited by L. Edel and L.H. Powers, OUP, 1987, p. 350.
12. There are 36 entries noting some of James's meetings with Mrs. Clifford in *The Pocket Diaries of Henry James*.
13. As note 9, item 57.
14. Leon Edel (ed.), *Henry James Letters*, Rupert Hart-Davis, vol. 4, February 24, 1896
15. Adeline Tintner, *Henry James's Legacy*, Louisiana State University Press, 1998, pp. 55-58.
16. A view communicated to the author by Leon Edel in 1996.
17. ibid, note to letter of August 9, 1899.
18. H. Montgomery Hyde, *Henry James at Home*, Methuen 1969, p.180.
19. As note 9, item 7.
20. Theodora Bosanquet, *Recollections of Henry James in his Later Years*. BBC Sound Archives, London 1956, Acc. 466 3R, Compiled and introduced by Michael Swan.
21. As note 10.
22. As note 9, item 73.
23. ibid, item 29.
24. ALS (one of 10), Percy Lubbock, Valehouse Collection.
25. As note 9, item 41.
26. ibid, item 42.
27. L. Edel (ed.), *Henry James Letters*, Harvard University Press 1984, vol. IV p. 606
28. Leslie Stephen, *The Mausoleum Book*, Clarendon Press 1977. p. 80.

29. Ann Oliver Bell (ed.), *The Diary of Virginia Woolf*, 5 Vols., Hogarth Press, vol. 1, p. 254.
30. As note 9, item 5.
31. As note 9, item 33
32. Theodora Bosanquet, *Personal Diary, 1914-1918*, unpub. Harvard University Library. The Bosanquet diary includes many references to Lucy Clifford.
33. Leon Edel, *The Life Of Henry James,* Rupert Hart-Davis 1972, vol. 5, p. 492
34. As note 9, item 81.
35. Theodora Bosanquet, *Diary Notes 1912-1916*, Archival Microfilm bMS Eng 1213.2(1), The Houghton Library, Harvard.

Trans-Atlantic Friendships:
Oliver Wendell Holmes Jr and J.R. Lowell

Early in 1875, Lucy and William met a young American from Harvard. His father was extremely well known and respected in America and in London, and he himself came to be popularly known as 'the greatest living American'. Lucy kept up a fifty-four year friendship with him. The young man was Oliver Wendell Holmes Jr. and it was partly through him that Lucy met another most distinguished Harvard scholar, James Russell Lowell. The story of these eminent men who moved in the highest social and academic circles on both sides of the Atlantic and who became very close to Lucy, gives some insight into the nature of her personality and the lasting qualities of her friendship.

In June 1866, when he was twenty-five, Oliver Wendell Holmes Jr. made his first visit to London. Five years earlier, while still a student at Harvard Law School, he had joined the Union Army to fight in the Civil War, seen violent action and been wounded three times. His father was Dr Oliver Wendell Holmes, the famous academic physician who was well known as the author of *The Autocrat at the Breakfast Table* and *The Professor at the Breakfast Table*. In London he received a warm welcome through his father's acquaintances and admirers, and his attractive personality and good looks soon made him sought after for himself. He already knew Leslie Stephen who had visited his parents in Boston in 1863. At that time Leslie Stephen had suggested that young Wendell might enjoy a climbing trip in Europe, and so, from the social round in London, he set out for the Swiss Alps. Oliver Wendell Holmes's diaries show how memorable the spectacular, and for him, sometimes terrifying, climbs had been. Frederick Pollock was one of this group of enthusiastic walkers and climbers who were later known as the 'Sunday Tramps' and, although the date of his actual meeting with Holmes is uncertain, the two men became firm friends. Fred Pollock, later wrote: 'There was no stage of acquaintance ripening into friendship; we understood each other from the first and were friends without more ado.' Oliver Wendell Holmes went on to become a Justice of the United States Supreme Court and is remembered as one of the greatest of all judicial figures. They pursued parallel brilliant careers on opposite sides of the Atlantic. The letters they exchanged over sixty-one years of friendship were eventually published in 1941, and elegantly mirror the intellectual and cultural activities of both men as well as providing a commentary on the political, judicial and social developments in the affairs of both nations over the years 1874 to 1932.[2]

After his alpine exploit, Holmes returned to the States. In 1874 he was back in London with his wife, Fanny. He immediately took up again with Leslie Stephen and the group of scientific naturalist thinkers who surrounded him. Holmes, although he had studied law, prided himself on holding a scientific attitude to all problems, and he relished the company of Herbert Spencer, G.H. Lewes, John Tyndall and Thomas Huxley. It was at this time that he met William Clifford who was then at University College and engaged to Lucy. The couples met together at a dinner party and Holmes noted that Lucy, like Fanny, was quick-witted and observant. On his next visit to London in 1882 he was alone. He met again with Leslie Stephen, now married to his second wife Julia. William Clifford was dead and Lucy was working hard to support herself and her two small daughters. Holmes was attracted by her intelligence, her independent attitude, and her bold sense of humour; the friendship that grew up between them lasted till her death. Sir John Pollock wrote of Holmes:

> he was, in the best sense of the word, a great ladies' man. He revelled in their society, his pen was stimulated by them, his conversation with them was as good as his letters to them, or better; but it was not all women who could please him . . . . He liked pretty girls, but what interested him still more were women of the world with enough brains and beauty to meet him on his own level.[3]

This distinguished, handsome and charming man was attracted to Lady Pollock and Lady Desborough, and an intensely loving relationship developed between him and Clare, Lady Castletown. He enjoyed flirtatious friendships both in America and in London and, although he was always loyal to his wife Fanny, the fact that she disliked public life and travel and had periods of ill health, provided him with the freedom to spend time with women friends. In London he always booked into Mackellar's hotel in Dover Street. *The Ladies' Athenaeum*, of which Lucy was a member, was also in Dover Street. It was of the summer of 1889, when Wendell Holmes was forty-eight and Lucy forty-three, that his biographer, Sheldon M. Novick, wrote: 'With Lucy Clifford, Holmes began a flirtation that became something more'[4]. Since she was a widow, she was an acceptable female companion and he loved to entertain her in his rooms at Mackellar's. To the end of his life he remembered moments of cosy intimacy with Lucy as they rattled round the streets and squares in hansom cabs – something they both loved to do. When he returned to London, again alone, in 1896 he took Lucy about on social calls, to dinners at the Savoy, and to the Indian Exhibition.[5] He also met again with his hometown friend and contemporary, Henry James, with whom he had a strangely uneven friendship. He returned to London in 1898 and was seen about so much with Lucy during his four-week stay at Mackellar's that Sheldon Novick remarks that they were said to be like an old married couple. Lucy by then was a well-known writer. She had adapted her short story: *The End of her Journey,* as a play: *The Likeness of the Night,* and she gave a private reading of it while Wendell was there. In it a barrister's meek, adoring wife commits suicide in order to leave him free to marry his mistress. The play later enjoyed a West End success. He visited London again in July 1901, took his rooms in Mackellar's and spent time with Lucy and other old friends. He was now Chief Justice on the Massachusetts

*Oliver Wendell Holmes, Jr. Over the years Holmes had given Lucy several photographs of himself. When in 1916 she received a photograph of this portrait she wrote to tell him how impressed her visitors were with it, but she teased him affectionately –
'The gown is so effective but oh my dear it is a pity that you are so ugly!'
The portrait is held in the Harvard Law School Art Collection.*

Supreme Judicial Court. He spent six summer weeks in London again in 1903, and was so warmly welcomed that he wrote: 'I feel twice the man I was after a visit to London.' He appreciated Lucy for her personality and lively mind and wrote that she was 'a brick with a heroic soul'.[6] In 1909 his wife Fanny accompanied him to Oxford where he received an honorary degree of Doctor of Laws. In 1913 he was seventy-two and made his last visit to London. War was threatening, many friends had died or were failing in health and Wendell Holmes was dismayed by this changed and sadder London. Lucy was still busy writing and entertaining her literary friends. She found it increasingly difficult together books published or her plays produced, but her friendship with 'dear OWH', as she called him, had survived the years and the separations, and she travelled with him to Southampton to see him board the *Lusitania* and sail out of her life. She wrote a parting note: 'What a good time we had – the best part of all this journey. Bless you – I shall think of you and watch the boat.'[7]

Holmes's intimate personal relationship with Lucy, which throws much light on the quality of her friendship, did not end with his departure on the *Lusitania*: they continued to write to each other as frequently as ever. He paid her the compliment of writing seriously about the legal problems he was called on to judge, telling her for example in 1911 about the difficult decisions in the Standard Oil and Tobacco disputes. Both of them wrote about the books they were reading, and Lucy sent him copies of her own publications and valued his comments. This man, one of whose biographies was titled *The Greatest American,* wrote to Lucy: 'I am always proud of your friendship.'[8]

The letters of the later years show us two friends in their seventies and early eighties, refusing to be old, refusing to let the fact that they would never meet again dim the warmth of friendship, sharing their critical enjoyment of books, and sustaining each other by keeping fresh the tender magic of their earlier affectionate association.

Seven years after her 1875 meeting with Oliver Wendell Holmes, Lucy met his compatriot, the famous poet, writer, editor and diplomat, James Russell Lowell. He and Leslie Stephen had been the closest of friends for twenty years and for Lucy there could have been no higher commendation. When they met, Lowell was sixty-four and Lucy was thirty-six. A brief look at his life and the events that brought him to live in London set the scene for his friendship with Lucy, which lasted till his death.

On 22 January, 1880, Rutherford Birchard Hayes, President of the United States, sent a dispatch to his Minister in Spain. It contained the surprise request that the Minister should transfer to the Court of St James' and become United States Minister in London. James Russell Lowell, then aged sixty-one and well known as a poet and writer but new to diplomacy, hesitated to accept the new posting. The Spanish appointment had been his first venture out of the publishing and academic worlds and he felt that the London post needed a more experienced diplomat. Born in Cambridge, Massachusetts in 1819, his professional career had been in publishing and editing and writing articles, essays and poems for various magazines including *The Atlantic Monthly* and the *North American Review*. Lowell is chiefly remembered for his many books, among them *My Study Windows* and *Among my Books,* and for his powerfully patriotic series of poems, the *Biglow Papers*. He had been an eloquently active pro-abolitionist though he was never a politician. His academic appointment in 1855 as Professor of Modern Languages and Literature at Harvard established him in the top rank of academic life. Honorary degrees at Oxford in 1873 and then at Cambridge in 1874 were accolades in his career. On retirement, Lowell and his wife travelled extensively in Europe. It was then proposed that he should represent America in Europe and he duly set sail for Spain to become United States Representative in Madrid. He remained there until the nomination to the Court of St James arrived Lowell was still something of an amateur for this high office. However, he was no stranger to England where he had plenty of good friends.

The story of the friendship between Lowell and Leslie Stephen is remarkable. Lowell writes that their first meeting, in Boston in 1863 when he was forty-four and Leslie Stephen thirty-one, was 'love at first sight'. He later elaborates the sentiment in this way: 'Whatever happens, my dear Stephen, nothing can shake or alter the hearty love I feel for you. I was going to say affection, but the Saxon word has the truer flavor.'[9] Leslie Stephen, in a letter to his mother, wrote that James Russell Lowell was 'one of the pleasantest men I ever met, and . . . has no humbug about him in any way'.[10] He felt that Lowell, through his *Biglow Papers*, was 'the creator and inspirer of that enthusiasm which in due time was destined to crush the gigantic curse of slavery'.

With pleasant anticipation of renewing an important friendship, Ambassador Lowell packed his bags in Madrid and set off on his mission to London and settled into the official residence in Lowndes Square. His wife was delicate and found social commitments a strain, but he enjoyed the wide range of duties and engagements that the Court of St James required of an Ambassador. He became a great favourite in London. It was soon chattered that Lowell had 'seen the inside of more country homes in England than any other American who ever lived'. He spent a lot of time with the family of Leslie Stephen both in London and at Talland House in St Ives and he met Lucy Clifford in the autumn of 1883. They met frequently while he was resident in London and a remarkable friendship developed. After he returned to live in Boston, the friendship was sustained by their correspondence and refreshed by his annual visits to London until his death in 1891. Lucy saved many of the letters he wrote to her,[11] and eleven of Lucy's letters to Lowell remain in the Houghton Library Lowell Collection. They wrote to each other in a bantering, affectionate, chattering style that suggests a very close understanding. During their friendship, Lowell gave Lucy copies of all the books and articles he had written. She sent him some of her books and invited him to comment on them. A paragraph from one of his biographies illustrates clearly why Lowell would find Lucy an attractive companion. He is described as:

> peculiarly dependent on the company of women, and he attracted to himself the wittiest and most responsive, for it was not so much the cushioned comfort that he looked for, as the cosiness of good fellowship and the intellectual equality which he sometimes found and always prized. He loved the generous natures with whom he had good converse, and his talk and letters went freely to these habitual dwellers in a world of honest sentiment.[12]

Both Lucy and Lowell displayed great sensitivity to the feelings of others. They may each also have recognised that they shared the vulnerability of actually needing to be liked. Indeed Lowell himself wrote:

> I would rather be loved than anything else in the world. I always thirst after affection, and depend more on the expression of it than is altogether wise.[13]

Lowell even felt secure enough with Lucy to sometimes drop his light-hearted banter and discuss, in his letters to her, the depressive streak in his nature.

He wrote at least four letters to Lucy in December 1884. He wrote on Christmas Day and Lucy's reply was written on New Year's Day from St Ives. She was spending the winter at Talland House as a guest of the Stephens and, as she so often does in her writings, gives a delightfully feminine impression of the occasion:

> for a couple of days the Fred Pollocks are staying and we shall talk philosophy and I shall try to look as if I understand it and liked it – tho' I didn't, I hate it, for Philosophy as far as I know only means trying to reconcile yourself to disagreeable things you can't help and I never do that. Well, on Saturday I am going to town to 26 Colville Road . . . . Come to see me soon, will you, to cheer me up? I shall be at home all day long on Sunday – from 11.30a.m. to 10 p.m. that means . . . . James

Gow has sent me his book on Greek Mathematics, pretty hard to read and much of it I am forced to leave alone, but all I understand of it is excellent, and I do love my old Greeks dearly . . . . We had a quite nice Xmas Day . . . the children dined late for the first time in their lives . . . . I carefully put on my best dinner-party looking frock, and my longest gloves and fancied myself a great deal with my biggest fan and we had a great many candles with rose-coloured shades, and flowers, and just as we were sitting down a band and 15 carol singers came and we gave them some money and they sang very old ditties in front of our window with fresh and not very much out-of-tune voices, and altogether the children – children with their hair dressed and tied back with white lace, both sitting bolt upright – thought they'd never known anything so grand in all their lives, and felt sorry for the poor Queen at her humble meal at Osborne . . . . After all 'making pretend' isn't so foolish as it seems sometimes. It helps one along many weary roads. But I am tired, and you must be too. It does me good to think of seeing all my friends again. I'll go and dream of it. Yours sincerely, Lucy Clifford.[14]

The pace of the letter-writing slowed down after 1885. In that year Lowell and his wife had a long vacation tour in Germany and Italy and then, on 19 February, 1885, Mrs Lowell died. Norton publishes part of his first letter to Lucy after his wife's death. He sent it to Talland House where she was again with the Stephens. How poetic and touching the first sentences must have been to Lucy, and how encouraging and flattering the remarks that follow:

March 19th 1885
Dear Mrs Clifford In trying to piece together the broken threads of my life again, the brightest naturally catch the eye first. I write only to say that I do not forget. Your letters are so complete in themselves that they can get on without an answer better than most. Could I write such nice ones I should do it for the mere fun of the thing as a game I could play without a partner – a kind of cup-and-ball, say. You round off your sentences so nicely and then, tap them into space and catch them so unerringly (if their own lightness ever lets them down) on the point of your pen that the game must be amusing enough in itself.[15]

Lowell's diplomatic appointment ended in 1885 and he returned to Boston.

He visited England each year and, in the summer of 1888, he is still 'flirting' with Lucy: 'I still have vigour enough to bring me with all speed to 26 Colville Street . . . when I hope the grey-blue eyes will show they are not sorry to see me.' And when asking her out, 'It's all right: I have got a chaperone who will reciprocally find shelter behind your skirts, so come, frock or no frock – I won't put on all my Orders!'[16]

His last visit to London was in 1889, and, although there were other relationships more important to him, it is clear that his friendship with Lucy was as warm as ever. In 1890 he wrote from *Elmwood,* his beautiful home in

Boston, that he would love to drop in to see Lucy again at Colville Road. But his hopes of another visit to England were never realised. He rallied from a serious illness, finished and published a new edition of his works, but died in 1891 aged seventy-two.

The last extant letter from Lucy to Lowell is in the Houghton Collection and is dated 11 May, 1890. She knew he was very seriously ill but cheered him up with chatter of her social life and amused him with the account of Rudyard Kipling standing godfather to her children's black kitten and 'solemnly christening it *Skuttles*'.

Lowell's life had been remarkably rich. He had experienced many losses – three of his children had died in infancy and his wife had died while still young – but throughout his life he had been surrounded by friends and admirers. When, in June 1885, after the death of President Garfield, he was recalled to the States, Queen Victoria commented that during her long reign 'no ambassador or minister had created such interest or won so much regard as Mr Lowell'.[17] He was held in such great respect that, when it became known that he might be recalled to the States, his academic friends and admirers proposed that he should be offered the Professorship of English Language and Literature at Oxford, but he did not take up the offer. Over the years he had given Lucy copies of all his books and writings. He encouraged her in her writing. It was he who urged her to write two of her books –*Love Letters of a Worldly Woman* and *A Modern Correspondence* – in the form of letters. Lucy's youthful personality and the half teasing, combative nature of their verbal exchanges – he loved, as he called it, to 'fence' with her – must have made him feel young. In spite of the difference in their ages the friendship was precious to both of them.

Notes

1. Mark de Wolfe Howe (ed.), *Holmes/Pollock Letters*, Harvard University Press, 1961.
2. ibid.
3. ibid, Introduction, p. xxviii.
4. Sheldon M. Novick, *Honorable Justice: The Life of Oliver Wendell Holmes Jr.*, Little Brown and Co., 1989, p. 188.
5. ibid, p. 208.
6. ALS (one of nineteen), Valehouse Collection.
7. ALS (one of four of Lucy's letters to O.W.H). Harvard Law School, O.W. Holmes Jr. Papers, MS Box 39.
8. As note 6.
9. C.E. Norton (ed.), *Letters of James Russell Lowell*, Harper and Brothers, 1894. Vol 2, p. 494.
10. Frederick Maitland (ed.), *Life and Letters of Leslie Stephen*, Duckworth, 1906, p.114.
11. There are 39 AL'sS, from James Russell Lowell to Lucy Clifford in the Valehouse Collection.
12. Horace E Scudder, *James Russell Lowell: A Biography*, Houghton Mifflin, 1901, p. 324.
13. C.E. Norton (ed.), *Letters of James Russell Lowell,* Harper and Brothers, New York, 1894, Vol 2, p. 75.
14. Eleven letters from Lucy Clifford to James Russell Lowell are held at the Houghton

Library, Harvard. This is from one dated New Year's Eve,1884.
15. ALS, Valehouse Collection.
16. ibid.
17. Ferris Greenslet, *James Russell Lowell,* Archibald Constable and Co., Boston and New York, 1905, p. 211.

Bloomsbury Connections:
Leslie Stephen and Virginia Woolf

The distinguished trio of academics, Sir Thomas Huxley, Sir Frederick Pollock and Sir Leslie Stephen had earlier taken it upon themselves to try to save William Clifford's life. They had advised, intervened, organised and financed the trips abroad in their endeavour to keep him alive, and they stood guard for Lucy and her daughters after his death. Leslie Stephen was particularly fond of Lucy and she was a frequent visitor at his London home. While on holiday in Cornwall, Leslie Stephen had come across Talland House in St. Ives and fallen in love with it. He bought the lease in 1881 and it was used as a family holiday retreat for thirteen successive years. Eighteen years later Henry James was to follow the same path when he found and fell in love with Lamb House in Rye. Each house was to exert strong influence upon its owner and occupants. The indelible impression of it on Leslie Stephen's daughter Virginia illuminates her book *To The Lighthouse*. Leslie wrote to Lucy in July 1884 to describe Talland House and to put it at her disposal:

My Dear Lucy,
    You asserted in your last note that I never wrote to you; in which there is some truth. So long as I am a dictionary-ridden animal [he was compiling the Dictionary of National Biography] I write to no one till I am free. And now I am free for three or four days, and show my natural disposition by immediately blossoming into correspondence. I have, however, only one thing really to say. We are here on a lovely blowing breezy day: the air is delicious – pure Atlantic breezes – not a germ in a cubic mile of it; you could not catch a disease without breaking all the laws of nature . . . and it is as soft as silk; it has a fresh sweet taste like new milk; and it is so clear that we can see thirty miles of coast as plainly as we see the back of Queen's Gate from our drawing-room window in London. We have a little garden, which is not much to boast of, and yet it is a dozen little gardens each full of romance for the children – lawns surrounded by flowering hedges, and intricate thickets of gooseberries and currants, and remote nooks of potatoes and peas, and high banks, down which you can slide in a sitting posture, and corners in which you come upon unexpected puppies – altogether a pocket-paradise with a sheltered cove of sand in easy reach (for 'Ginia even) just below . . .
. you must come here when we go (or rather before we go), and take up your abode with Ethel and Alice. Seriously, I feel that it is a sin to

leave the place to itself for so long a time as our absence, and it would be an unadulterated pleasure to think of you and your little ones getting some good out of it. [1]

Lucy visited and then spent November and December of that year there. It was there that she began writing what was to be her best-seller, *Mrs Keith's Crime*. Leslie wrote wishing that he could join her and guide her on some walks along the St Ives lanes but complained that he was beset with the labour of editing what he called 'that Damned dictionary' and 'my accursed burden'. He wrote of it: 'It is about my bed and about my paths and spies out all my ways as the psalmist puts it; . . . I must set to work to-night, having, however, refreshed myself by talking to you and getting an imaginary breath of St Ives.'[2]

The letters that Leslie Stephen wrote to Lucy and William reflect the easy nature of their relationship and bear out what Lucy later told Frederick Maitland, as he prepared his biography of Stephen, about this man with a reputation for anti-social behaviour and grumpiness:

> He was simply delightful with children. Sometimes in his house he went down on all fours and gave two of his children and my two, all tiny, a ride on his back, all of them at once. One of my daughters was very ill when she was about four years old. Leslie came with some sheets of white paper and scissors, and asked if he might go upstairs and sit with her for little while . . . . They were soon holding an animated conversation, while he, with apparently no thought about it, cut out a series of animals – beautifully formed giraffes, bears, goats, dogs, anything – that filled her with delight . . . no one who had not children could realise what he could be to them.[3]

In 1885 he sent from St Ives the most delightful letter with a charming illustration from his dog Rob to Lucy's dog Spot. The famous lighthouse is shown in the background. Rob reports that his master is 'grown quite fat on Cornish cream. I am going to take him to the dictionary and set him to work. This afternoon Nessa and Jo and Ginia and Adrian are all very well.'[4] The families shared many friends and James Russell Lowell, who was so close to Lucy, was godfather to Virginia, as Leslie Stephen expressed it, 'in a quasi-sponsorial relation'. He gave her a live bird in a cage and the other Stephen children were jealous. It was sad that the childhood friendships between Virginia and Vanessa and Lucy's two girls did not last into adulthood.

Lucy would fuss over Leslie Stephen and warn him against overwork. He replied:

> The feminine mind has every merit; but it is haunted by the strange illusion that men overwork themselves. I have never in my life worked hard, except when I was taking my degree; and I grow steadily lazier as I grow older. The only reason why I ever get anything done is that I do not waste time in the vain effort to make myself agreeable.

Leslie Stephen's dog Rob delivers a letter to Lucy's dog Spot, showing Stephen's artistic skills in an illustration to a letter. The lighthouse behind was later immortalised in his daughter Virginia's book To The Lighthouse.

She left many humanising vignettes of this man who so often was remembered as difficult and remote. Ethel and Margaret played with the Stephen children and Leslie Stephen, joining a game of Reversi with them, would sulk because of Ethel Clifford. 'She would beat me, confound her, I never like being beaten.'[6] His colourful outbursts included one when he was still in orders, to a man who swore in his presence, 'Damn your soul, Sir! Don't you know that I'm a parson?' Certainly, Leslie Stephen was not an easy man. Thomas Hardy, visiting Stephen in 1886, noted in his diary that his behaviour was: 'just the same or worse; dying to express sympathy, but, as if suffering under some terrible curse which prevents him from saying any but caustic things, showing antipathy instead'.[7] This was the year that Stephen collapsed from overwork and was ordered to take a complete rest. Lucy was staying at Clarens, on Lake Geneva, and Leslie Stephen went to stay with her there. Together they set off to Zermatt for the New Year. Lucy reported on an exchange they had with a fellow hotel guest:

> I thought I was in some fashion paying back the parson's kindness when I invited him to meet Leslie. But that parson, bless you, had never heard of Leslie and to my horror I heard him say, 'Oh yes Mr Stephen, I like some Radicals, I assure you. Now there was Mr Fawcett – did you ever come across him?' Leslie gave a grumpy 'Yes.' I settled myself at the tea table and said firmly, 'Oh, but you forget that Mr Stephen is Mr Fawcett's biographer.' 'Indeed?' said the parson in a benevolent voice, 'Do you do much writing?' Leslie gave a grunt, and I put in, 'Mr Stephen is editing a little Dictionary,' and as that produced no response I added, 'The Dictionary of National Biography,' for I felt bound to boast a little

when I saw my poor friend being sat upon. 'Indeed,' beamed the parson, 'It must be very interesting.' The thread of conversation wandered to the scenery, 'There are some beautiful walks about here,' said the parson, and he asked the shepherd of the Sunday Tramps in a kind of patronising manner, 'Do you care for walking Mr Stephen?' Leslie was much too squashed even to answer, and I again struggled to the rescue with, 'Mr Stephen is a great walker and a little stroll of twenty miles before breakfast is nothing to him.' 'Indeed!' said the parson incredulously. 'Ever do any climbing?'[8]

Lucy gleefully reported that a great impression was later made on the guests when the famous Alpine guide, Melchior, arrived with ropes and axe and set off with Leslie for the mountain peaks. There was a reason perhaps for the clergyman's misconception, for Leslie Stephen's appearance did not suggest athleticism. One of the endearing and distinguishing aspects of his appearance was that he paid little regard to clothing. He always wore his ordinary day-to-day clothes plus hiking boots for climbs and long tramps across the moors. While he was at Zermatt, Ethel and Margaret sent him a present. He wrote a charming letter of thanks and enclosed delightful drawings of himself as a bear. He really enjoyed the time in Zermatt and wrote to Lucy afterwards that the solitude enabled him to reduce himself to a state of 'intellectual annihilation . . . . I often sit like Brahma and contemplate the lowest button on my waistcoat with satisfaction. It is in danger of coming off after meal times.'[9]

The Sunday Tramps was the élitist band of walker-philosophers formed by Stephen in 1879 and led by him. He very much enjoyed marshalling the group and masterminding the sometimes complicated planning of the expeditions. Lucy wrote to Vanessa after Leslie's death. She described how on one occasion The Tramps had broken off their walk to dine with William Tyndall at Box Hill:

Leslie sat at the head of the table and did his autocracy with an occasional grunt or groan if someone said something absurd, but we all knew him to be in his glory. Afterwards we had coffee in Tyndall's study upstairs and then Leslie looked round and said to himself so that all who were near him could hear, 'I must sweep these creatures off' and five minutes later The Tramps were striding down the road two by two, Leslie's tall figure erect and unfaltering, a pace or two in front.[10]

The society flourished for fifteen years, walking every other Sunday for eight months of the year till, as Maitland put it, 'the deceitfulness of golf and the vanity of bicycles distracted some of those who had been consistent walkers'.

In 1887 at Lucy's lodgings in Clarens, Leslie Stephen met her friend Olive Schreiner. He reported her as 'a pretty, black-eyed, tiny woman of 25 or so who has written *The Story of an African Farm* . . . she is clever, but, I should guess, hard and conceited'. As he got to know her better, he decided 'she disapproves of marriage and thinks that everybody should be free to drop everyone else – I should drop *her* like a hot potato'.[11] Lucy was loyal to Schreiner, who suffered from poor health, and she

The inscription reflects Leslie Stephen himself working on his Dictionary of National Biography. *It reads, 'This bear has been reading a dictionary and is quite tired, poor bear! But he can put in a marker and read no more.'*

'This bear has just made a good joke - So he is writing it down in his notebook that he may not forget it.'

listened sympathetically to the problems of her relationship with the mathematician Karl Pearson, but she would not have been supportive of her progressive views on sexual freedom. Leslie Stephen was close enough to Lucy to be able to show his feelings and, in his letters to her, gives insights of his struggle against depression. When Stella, the child of his wife Julia's first marriage, died he wrote:

My Dear Lucy,
　. . . We are a sad party. Ginia is still nervous but getting better. J. Hill [Stella's husband] has been here for a Sunday or two. He takes his calamity very well; quietly and bravely: – but he is silent, I think

by nature and said very little . . . . It still requires a little effort of the imagination for me to realise that anybody can be more concerned in the loss of Stella than I am . . . I have less strength left than they, and feel how little I have to look forward to . . . . I find it harder than I ought as a sensible man to accept the fact that the good things in life lie all behind me . . . . I must not go on croaking like this. I do it because it is a relief to me; and, at intervals I try to make up my mind to be a little more cheerful. Generally I relieve myself by getting on to my accursed and yet blessed treadmill [the *Dictionary of National Biography*] which makes me into a machine and stops my whining – I am not really unhappy so long as I can 'lie low'. Thoby is rejoicing over the capture of a butterfly that is neither a 'Clifton Blue' or a 'Chalkhill Blue' but something between the two. And, though I don't see why that should make him ecstatic, I am glad to see that it seems to satisfy him in every fibre.[12]

Lucy and Leslie corresponded and met regularly. In 1900 he wrote: 'Caught a chill and so did not call. I saw Henry James to my great pleasure. He is one of the few *men* who manage to keep up an interest in me. I am richer in female friends!'[13] Leslie Stephen recorded his affection and gratitude for Lucy's sympathetic relationship with Julia. He commented that 'few people loved my darling more'. Towards the end of his life, especially after reading in Froud's biography of the irritable intolerance shown by Carlyle to those close to him, he was fearful, perhaps with reason, that he had never fully demonstrated his affection and love for his family. He was therefore delighted to record the following report passed to him by Lucy:

Henry James, the novelist, who had always loved Julia, was speaking to Mrs Clifford before her [Julia's] fatal last days. He said 'Good God, how that man adores her!' I require no proof of my adoration, but in my morbid state it was delightful to hear that I made it evident to my friends. If they could see it, did she not know it? Mrs Clifford told me more to the same effect, and spoke so gently and cordially and sincerely that I am full of gratitude to her . . . .[14]

Leslie Stephen became ill with cancer and, early in 1904, he died. To the end he kept his friendship with Lucy, and the affection he held for her is shown in this letter to Lucy from her close friend Lady Ritchie, who was Thackeray's daughter and herself a novelist. Her sister, Minnie, had been Leslie Stephen's first wife. She wrote just after the funeral:

Dear Old Friend, Your letter is just a feeling – not words and it is one I share and can feel happy over as well as unhappy because that kind fire at which I have warmed myself for so many years is gone out. When I went to Highgate and saw Julia's name and Stella's and the flowers from you and me and how many others who loved him, lying in the mist – I felt as you do, not that Leslie was dead but that the fire was warming one . . . . It was like a farewell that day when he said 'I want to see Lucy Clifford. I must see her, whoever comes.' The next day he was sadder and so weary of life except when he went back to quite, quite early days and spoke of his sister as if they were boy and girl again.[15]

Maitland writes that on Sunday, 21 February, 1904, Stephen had 'talked about books and people to his last visitor, a lady. Early the next morning he died.' Remembering his wish to see Lucy above all others, this last visitor could well have been her. After his death she was active in canvassing for contributions for a memorial to him, and a framed copy of the memorial photogravure of him hung prominently in her sitting room in Chilworth Street. Lucy, ever grateful for his love of her husband, had been the most supportive of his companions, providing a haven of sympathy and comfort to him in his years of sickness and loneliness.

Before their father's decline in health, his children and step-children had been involved in the sort of social round that Virginia and Vanessa particularly dreaded – formal parties at which they felt so constrained and shy that they often remained silent all evening:

> On Sunday afternoons they did a round of 'at homes', exchanging one matron's drawing room filled with blushing marriageable daughters for another, Vanessa listening to Lady Kay Shuttleworth on the beauty of sunsets, or to Mrs Humphry Ward, already adverse to feminism and later to be a vociferous opponent of the Suffragettes. She encountered Mrs W.K. Clifford (author of the best-seller *Mrs Keith's Crime*) and her daughter Ethel who affected the pose and dress of the 'aesthetic' movement but who looked even less like the Burne-Jones ideal than her novelist-mother whose teeth were awry and who delighted in gossip about the literary world.[16]

Their father's last illness had heavily affected the lives of his daughters:

> For two years this meant more than ever a life of seclusion from our generation. Visits from his friends had to be arranged and carefully fitted in so that he had the right number each day. This was difficult and exacting for we had to be prepared to entertain Mrs W.K. Clifford and old C.B. Clarke while they waited to replace Mr Haldane and Sir Alfred Lyall in his room. There was of course no telephone and innumerable arrangements had to be made by letter or otherwise. And so no time or inclination was left for other society had it been possible.[17]

From letters and diaries it is clear that Lucy Clifford and others of her age and background had come to represent to the Stephen girls, Virginia and Vanessa, the oppressive, stuffy, conformist society from which they longed to escape. After their father's death they took up their freedoms. The 'Bloomsbury' group evolved, centred on Gordon Square where Virginia and her brother Adrian set up house. Vanessa married Clive Bell in 1907 and began to establish herself as a painter. Virginia married Leonard Woolf in 1912 and dedicated herself to writing, and the generation gap widened even more. The Bloomsbury group's commitment was to destroying 'all barriers of reticence and reserve'. Virginia showed her feelings when she wrote to Lytton Strachey, 'I was suffocated in Lady Pollock's drawing room this afternoon – you never saw such a sight. . . . O how those old women spoil my life. Think of the embraces of Mrs Clifford and Aunt Anny [Lady Ritchie]!'[18] The controversial, highly intellectual, challenging, sexually free behaviour of the group, centred round Leslie Stephen's children, was light years away from the scene set by their parents and their parents' friends. Their paths, however, did sometimes cross. In January of 1920, Leonard and Virginia accepted an invitation to tea with Lucy. On their return home Virginia made this entry in her diary:

> Oh dear, though all this talk of novels is all turned sour and brackish by a visit to Mrs Clifford. She must have supplied herself with false teeth since I saw her twenty years ago and her hair frizzed out is surely browned by art; but she remains otherwise the same – large codfish eyes – the whole figure of the nineties – black velvet – morbid – intense – jolly, vulgar – a hack to her finger-tips, with a dash of the stage – 'dear' 'my dear boy.' – 'Did you know Leonard, that I was married for only three years and then my husband died and left me with two babies and not a penny – so I had to work – oh yes, I worked and sold the furniture often, but I have never borrowed.' However, the pathetic is not her line. She talks to fill up space; but if I could reproduce the talk of money, royalties, editors and reviews, I should think myself a novelist; and the picture might serve me for a warning. I think that one may assume it to be more of a product of the 90ties than of our age. Again, having years ago made a success she's been pulling the wires to engineer another ever since and has grown callous in the process. Her poor old lips pout for a bit of butter, but margarine will do. She keeps her private and very rancid supply on some of those little tables that those distressing rooms are lumbered with (a wooden black cat on the clock, and little carved animals under it). She has a review of herself in *The Bookman* and a portrait, and a paper of quotations about Miss Fingal. I assure you I can hardly write this down. Moreover, I had a feeling that in these circles people do each other good turns and when she proposed to make my fortune in America, I'm afraid a review in *The Times* is supposed to be the equivalent. Brave, I suppose, with vitality and pluck – but oh the sight of the dirty quills and the scored blotting paper and her hands and nails not very clean either – and money and reviews, proofs, helping hands, slatings – what an atmosphere of rancid cabbage and old clothes stewing in their old water! **We** went away laden with two of the heap of flaring books – 'Are you going to take my mangy books! To tell you the truth I'm in debt' Yes, but was that why we were asked to tea? Not altogether, I suppose, but, partly; sub-consciously.[19]

Virginia's entry in her diary struck where an ageing woman would be most vulnerable – her appearance and the standard of her hospitality. Mercifully, Lucy was long dead before the diaries and letters were published through the

Hogarth Press. Of course, a personal diary is a private document, and there was no reason for Virginia to monitor what she wrote. She could, and did, destroy with her pen and Lucy was not alone on the receiving end of it. Virginia mocked her cousin Emma Vaughan for 'a blankness in the eye which is precisely that of a toad glutted with large moths'. She stated that Katherine Mansfield 'stinks like a civet-cat' that H. G. Wells had the 'cheeks and jowls of a butcher' and was 'arrogant, lustful, bullying'. Lady Arthur Russell, Bertrand's mother, was, in Virginia's diary 'a rude tyrannical old woman with a bloodstained complexion and the manners of a turkey cock'. But throughout the many volumes of letters and diaries no destructive description is so detailed as the one she constructed for Lucy Clifford. She had every right to practise the detached observation that marks her as one of the greatest of all writers. But why so bitter an attack on Lucy? There was worse to come.

Later that year, after Lucy had been one of a group of guests at Gordon Square, Virginia wrote to Vanessa. She began with family news and chat and then continued:

> This however you skip; and want to know all about Mrs Clifford – who was, indeed, all that you've ever imagined her to be – wattled all down her neck like some oriental Turkey, and with a mouth opening and shutting like an old leather bag, or the private parts of a large cow. Her stories were magnificent – about Ruddy Kipling's wedding – and how [the publisher William] Heinemann came too late with a bouquet – and picture Sundays [the Sunday before 'sending-in day' at the Royal Academy when artists invited friends to see their work] and how Professor Sylvester suddenly sank on his knees on the black hearth rug and recited his sonnets to her, with Ethel [her daughter] at her breast – and then 'yes, a great many young men have come to me in their troubles, for a lone woman who has once been married and considers herself married still, is a great help to young men – and they all know that Lucy Clifford never gives away a secret and never thinks harm of a friend' – upon which we all fell silent and almost sobbed – This was to explain why she wouldn't write her memoirs. Moreover, she moves about all in black, lurching like a black beetle that's lost a front leg, so I suppose I shall have to review her novel after all – her courage and fertility move my heart to tears. Mrs Ward [Mrs Humphry Ward who died March 24 1920] is dead; less notice has been taken of it than you would think. She was escorted by constables to the grave, and old Lucy Clifford was there, too, of course.[20]

In 1923, Virginia wrote to Roger Fry:

> Can you advise me how to acquire the social manner – neither cold nor hot? When I go to these tea parties, they all seem like people enchanted, and chained to a particular patch of the carpet, which they can't cross for fear of death – But you know it of old. What irritates me is to see

– anybody, Mrs Clifford it may be – possessed of a sense which I have not. And I believe – but here you mightn't agree – that it is one essential for a writer. I think Proust had it.[21]

It is impossible that Virginia could have been jealous of Lucy, but she was always in some way uncomfortable about her as she showed in a letter to Clive Bell: 'On the way I ran into Mrs Clifford and Turkey [Margaret Clifford] – a curious and painful encounter.' And, in her diary, 'I'm exposed to the hanging lip and clamorous vanity of Lucy Clifford today: she has an article on George Eliot which she wrote for a special fee (that is where I shall end up if I don't take care – talking always of 'fees') for the *Nineteenth Century*.'[22] It seems almost as if, by enunciating the particular details of her aversions to Lucy's literary achievements and attitudes and lifestyle, Virginia was indicting the whole atmosphere that the Bloomsbury movement sought to escape from and would in the end destroy.

Eight years later Clive Bell, reports how Vanessa reacted to the news of Lucy's death. The Bells were at their property at Cassis in the South of France:

> [Vanessa] was almost beside herself with excitement when Mrs Clifford died: in fact she was preparing to drive the Citröen into Marseilles for the next day's *Times* in the hope that a correspondent would write – as of course he did. The ghoulish strain in the Stephens is very strong; in Vanessa it is the only definite literary heritage – except of course for her admirable prose-style.[23]

Virginia had written to Vanessa a few days earlier, 'Mrs Clifford is dead, and I've written a wreath-like letter to Ethel. She is buried today, and I can't go.' The next month Virginia again wrote to Vanessa from Monk's House in Sussex:

> How could I go to Lucy's funeral, seeing that I was here? All that remains of her in my mind is a cow's black blubbering cunt: why that image persists I know not. Nor have I heard from Ethel. My letter was not from the heart; it rang, as they say, hollow.[24]

This was, of course, a private letter, but Virginia's gratuitous obscenity is still shocking. She was not to know that her letters would later become public property, but it is not pleasant to think of her using such words against an elderly woman who had been a good friend to her father. Lucy held benevolent memories of the family and had preserved in her collection the following most gracious letter from Virginia:

> My Dear Mrs Clifford,
> I am sending you, with my love, a little book of my father's which we have just brought out, thinking you may like to have it. I only wish it were nicer to look at, but owing to our going away, the printers have made it look rather cheap and nasty, I'm afraid. But I like the book itself, and I think you may, too.
> Yours affectionately, Virginia Woolf[25]

The gift may have been in some way mandatory since on 12 May, 1924, Virginia had received this letter:

> My dear,
> Would you like these rough and hasty notes set down when F.W. Maitland asked me for letters and anything I could tell him about your father? Some of the letters he used; but there is no reason why you should not use them again, if you wish. Anyhow they may interest you. I should love to see you both again (I sometimes read you in the *Nation* and *Athenaeum*.) Could not you come to tea one day? But telephone me first, I should hate to miss you. I read, and liked so much, your notice of Hester's book about her mother. I visit you often in my thoughts, dear – and others that we both remember.
> Affectionately yours,
> Lucy Clifford.[26]

Virginia had few kind words for Lucy's daughter Lady Dilke. In 1927 she wrote to Vita Sackville-West:

> My no-clothes dodge is working admirably. I was rung up yesterday by a woman called Lady Dilke who wanted me to give the Femina Vie Heureuse prize to a Frenchman. So I said I couldn't. And she said you must. So I said I wouldn't and she said I should. So it struck me, well I'll say I have no clothes. At which she paled and withered, and cried off instantly: it was to be at Claridges in May. Also I think I've got out of lunching with her on the same plea. It's true, too. Never shall I lunch with Lady Dilke. Never shall I give a prize to a Frenchman. (And by the way, for your information she said something about giving me . . . the prize and I blushed all over, holding the telephone, with shame and ignominy. This is true. Snobbish? No, instinctive, right.)[27]

In a letter to Vanessa the next month she wrote:

> Then a voice on the telephone recalled me thirty years at a leap the other day. Something mincing, powdered, affected, vulgar, effusive, fawning, – who do you think? Ethel Dilke! She wanted me to lunch with her and give a prize to a Frenchman. But I promptly invited her here for a date unknown and so shall escape – unless you insist on a rapprochement . . . . Yrs B[28]

In fact Virginia was awarded the Femina Vie Heureuse prize the following year for *To the Lighthouse*.

It is not easy to understand the vehemence of Virginia's words and feelings about Lucy Clifford. A clue may perhaps be found in her writings. Describing the climactic boat-trip in *To the Lighthouse*, Virginia, through the eyes of Cam, views with scorn her father, fictionalised as Mr Ramsay, sitting in the boat savouring his self-image as 'a desolate man, widowed, bereft;' dreaming of 'the exquisite pleasure women's sympathy was to him'. Perhaps it was her father's steadfast affection for Lucy and the sympathy she offered him that was offensive to Virginia; perhaps she felt that he was himself diminished by seeking, through Lucy, that 'exquisite pleasure' after his adored second wife, Julia, – Virginia and Vanessa's mother – was dead. Such perceptions may have added the destructive dimension to what could have been an understandable aversion to an elderly, rather old-fashioned woman. There is no evidence that Virginia's dislike of Lucy and Ethel was in any way mutual; indeed, Lucy was always benevolent to the Stephen family. But, also, she had time, before her death, to edit all her papers and to make sure that there should be no possible embarrassment or hurt from them. Her daughter Margaret sifted through them too, and no unkind personal reference has been found in any of them. By contrast, Virginia's literary standing has ensured that not even her privately recorded writings, painful though they might be to others, are now concealed. Thus her wounding words about Lucy Clifford remain, and they provide a singular insight into our perception of both women.

Notes

1. F.W. Maitland, *Life and Letters of Leslie Stephen*, Duckworth and Co. 1906, p. 384.
2. There are 22 letters from Leslie Stephen to Lucy Clifford in the Valehouse Collection. Parts of four of them are quoted in F.W. Maitland's biography.
3. As note 1, p. 477.
4. ALS, Valehouse Collection.
5. As note 1, p. 385.
6. Noel Annan, *Leslie Stephen: The Godless Victorian*, Weidenfeld and Nicolson, 1948, p. 109 (there is much of interest about William Clifford in this volume.)
7. As above p. 143.
8. As note 6, p. 97.
9. ALS, Valehouse Collection.
10. As note 6, p. 98.
11. As note 6, p. 110.
12. As note 2.
13. ibid.
14. Leslie Stephen, *The Mausoleum Book*, Clarendon Press 1977, p. 80.
15. ALS (one of 4), Valehouse Collection.
16. Frances Spalding, *Vanessa Bell*, Harvest USA, 1983, p. 27.
17. ibid, p. 38.
18. N. Nicolson (ed.), *The Letters of Virginia Woolf*, 1888-1941, Hogarth Press 1975-1980, Vol 1, item 458, letter to Lytton Strachey, 1908.
19. A.O. Bell (ed.), *The Diary of Virginia Woolf*, Hogarth Press, 1977-1984, Vol 2, entry for 24 January 1920.
20. As note 18, Vol. 2, item 1126.
21. ibid, Vol. 3, item 1390.
22. As note 19, Vol 3, entry for 23 July 1927.
23. As note 16, p. 231.
24. As note 18, Vol 4. item 2026.

25. ALS, Valehouse Collection. This letter has not been printed before and is undated, but must be from 1924. The book referred to is *Some Early Impressions* by Sir Leslie Stephen.
26. ALS, Monk's House Papers, University of Sussex Manuscripts Department.
27. As note 18, Vol 3, item 1722.
28. ibid, item 1732.

London Friends at the Turn of the Century

For the first eighteen years after her husband's death, Lucy resided at what had been their home together, 26 Colville Road – north of Kensington Gardens and between Notting Hill Gate and Bayswater. She lived frugally with one servant to cook and clean, and a young nursemaid to help with her two daughters. In 1897 she moved next door to number 27. It was not a highly fashionable area, but had a quiet style about it. Lucy always gave her address as Hyde Park W. She instructed her friend, the writer Rhoda Broughton, on her first visit for tea, to 'Go to Notting Hill Station and walk the ten minutes to my house or, if you don't disdain the humble omnibus it comes from the station to my door in four minutes' she promised her that 'Henry James and Kenneth Graham and two or three others – all old friends of yours will be there.'[1] Rhoda Broughton was four years older than Lucy. Her books – considered to be 'sensational' because they depicted women in revolt against traditional social restrictions – sold very well. She wrote chatty letters to Lucy about her work and the books she was reading and was open about her financial problems. Lucy saved fifteen of her letters. They were able to make friendly criticisms about each other's work and in one of her letters Rhoda took Lucy to task for allowing 'contemptible male things' to have too much influence in one of her books.[2]

Lucy deliberately set out to 'garnish her salon' by inviting people – especially writers – who she thought would be interesting or whose books she would like to review for *The Standard*. Violet Paget, who wrote as Vernon Lee, had, after her first invitation in 1881, unkindly noted in a letter to her mother that: 'Mrs Clifford lives at the world's end and the visit cost me five shillings, certainly more than it was worth.' Vernon Lee was famous for her sharp tongue and outspoken comments, and she was critical of almost everyone she met that afternoon. She wrote of:

> This little apish creature, grimacing with a huge mouth, talking atheism sixteen to a dozen, and regretting that modern science could not revive the belief in hell in order to make people attend his lectures.

It was Sir Thomas Huxley – President of the Royal Society! She saw Lucy as, 'an agreeable sort of intellectual Lilah . . . with extremely bad manners in the way of talking across her visitors and so forth'. Karl Blind's son Rudolf she thought 'an awful little snob' and Leslie Stephen she reported as:

> A tall sort of solemn, scraggy, lantern-jawed Rubens type, who looked hideously shy and sat in complete silence for half an hour. On my taking

my departure he shambled forward and stammered inaudibly that he was sorry he had had no opportunity of speaking to me! Had I come to England to extend my literary connexions, or to enjoy great fame, I think I might go and hang myself on the first peg I met.

Her sharp words were somewhat softened after the next meeting:

Yesterday I called on Mrs Clifford and found her alone. When she is alone she is interesting and cordial. Her intolerable self-engrossment and ill breeding are counterbalanced by a kind of selfish expansiveness and expansive helpfulness, which are pleasant . . . besides, she has a great feeling and talent, and her extraordinary novels show it.

She came to greatly respect Leslie Stephen, and later, after Lucy had given a party in her honour, she wrote, 'I see daily more and more, that if any set claims me, it is the Clifford one.' Lucy introduced her to Mark Pattison, Henry James, Theo Watts, Norman MacColl, Frederick Macmillan, John Collier and many others, including Sir Frederick Pollock who was 'very civil' about her book *Euphorion*. She thought it touchingly odd that, in 1884, when she presented a copy of the book to Lucy, she insisted that she dedicate it 'To W.K. and Lucy Clifford' as if her husband were still alive.[6] Some of Lucy's other women writer friends had assumed masculine pen names to help them get their work published. Among them was Mary St Leger Harrison, whose pen name was Lucas Malet. Mrs Pearl Mary-Teresa Craigie, who wrote as John Oliver Hobbes, called Lucy 'the kindest of friends' who had 'encouraged her from the beginning'. Olive Schreiner, who wrote her best-known book: *The Story of an African Farm* under the name Ralph Iron, was very close to Lucy. She had a romantic involvement with Karl Pearson, who edited and completed William Clifford's *Common Sense of the Exact Sciences,* and Lucy was the friend she turned to for understanding when that relationship broke down. Lucy's own friendship with Karl Pearson was a long and close one. He saved forty-eight of her friendly chatty letters to him and they are held in the University College, London, Archives. She wrote to him from holidays and visits abroad, and she kept him up to date with news of Ethel and Margaret. In 1925 she tried to enlist his help in getting the Highgate Cemetery authorities to clear a path to William's grave. She was kind to him when he was ill and in a note to her he wrote: 'the tongues will not be wasted, the jelly was excellent. As for the rug, it is fit for a Pasha! Always yours, Karl.'[7]

William's friends continued to visit Lucy, and among them were the Blinds. Karl Blind, the political writer, had been exiled from Germany for his revolutionary activities. The Cliffords shared his adopted daughter Matilde's admiration for the Italian republican leader, Giuseppe Mazzini. Both William and Lucy were heavily influenced by his philosophy and each of them frequently referred to Mazzini in their writings. Lucy was 'darling Lucy' to Matilde who was then, in 1883, engaged on the very first biography of George Eliot. Lucy helped her by giving her letters which George Eliot had written to William.

Mary Cholmondeley, whose book *Red Pottage* was a great success in 1889, Mrs Braddon, May Sinclair, and Alice Meynell, the poet and essayist, were also among Lucy's close women friends. These women leavened the mainly male attendance at her tea parties. As well as meeting her friends, Lucy corresponded with them and saved many of the letters she received from them.

Mrs Humphry Ward was five years younger than Lucy and had her great success with *Robert Elsmere* in 1888. Mrs Ward sent her a copy of her 1896 publication *Sir George Tressady* and Lucy reviewed it in *The Standard*. Lucy wrote to her, 'Henry James is coming in tomorrow and we shall discuss your book. The F. Macmillans came in full of it and Leslie Stephen too, but he had only just got it and not read it yet;' she ends 'I rejoice in your success'. Lucy was 'Dearest Lucy' to Mrs Humphry Ward who signed herself as 'your ever loving friend, Mary A. Ward'. Both women were able to comment freely about each other's writing. The Humphry Wards were guests at Lucy's daughter Ethel's wedding in 1905 and Mary Ward offered to bring flowers for the event from the gardens at her country home, *Stocks*. In 1908 she wrote to Lucy 'you are the most generous and encouraging of friends'. She pressed Lucy and Ethel to visit, explaining that 'The motor will come to your door – you need only put yourselves inside and a bag on top and in one hour and three quarters you would be at the door and you know what sort of a welcome you would get there.' In 1914, she thanked Lucy saying 'You always write me generous and interesting things about my work and I love your letters.' In 1919, Mary wrote to Lucy wishing her success with her latest 'extremely fresh and interesting book', *Miss Fingal,* saying, 'If ever anyone deserved it, it is you – you dear hard worker and generous friend.'[8] A year later she was dead, and her husband wrote to Lucy mourning his wife and recognising Lucy as one of their truest friends.

Lucy was also gentle with her friends, but when a scandal arose because of the behaviour of two of them she was not afraid to write and boldly express her views. The friends were Violet Hunt and Ford Madox Ford who had run off together, and, having gone through a form of marriage abroad, intended to return to London and face the scandal that was buzzing around them. Violet wrote to Lucy for support. Lucy firmly advised her to stay abroad until the whole affair had been forgotten. She advised them they should prepare a statement to declare that:

> Ford divorced his wife on the — of the —. That the citation was served on her on such a date at such a place – give time, places above all things.
>
> The divorce decree was procured at such a court or before such a judge – give dates and places.
>
> On such a date you two went through the ceremony of marriage at such a place, believing it, as you now do, to be legal.[9]

She added a postscript 'For goodness sake, dear woman, set forth in plain clear words the facts, and say you do so, not for your own sake, but for the sake of your friends who do believe in you'. The Fords decided to ignore the scandal and return to London. Henry James broke his friendship with Violet but Lucy, and many others, remained loyal to the couple.

Arnold Bennett, who was twenty years younger than Lucy, had become friendly with her in the early 1900s and he much admired her book *Aunt Anne*. Like Lucy and many other British writers, he had discovered that a hotel in Vevey for 13 Swiss Francs a day was cheaper than staying in London and at least one of his books was written there. After Henry James's death, Bennett wrote to Lucy: 'I knew very little of him personally. He once asked me at the Reform Club if he might join my party there. I blushed. I really did. He was most amusing, sinister, and critical.'[10] In 1914, Lucy sent Bennett her play *Thomas and the Princess*. It had been published in her 1909 collection *Three Plays* but had not found a producer. Arnold Bennett made several suggestions to improve the construction of the play and suggested two possible producers. He also gave her a warning and some dubious advice:

> I attach little importance to my recommendation, or to anyone's, and I fancy that you yourself would be more persuasive than me – considerably. *But* from my knowledge of managerial minds, I am convinced that if this play were offered to either of these men in its book form, with its author's signature, its chances would be diminished. I should suggest that you have the play typewritten and make a purely temporary change in the title, and leave out the author's name. You need have no fear that the managers will discover the trick. They won't. They wouldn't recognise Measure for Measure if the title was altered. They read absolutely nothing in book form. . . . All this is merely a suggestion, my dear lady . . . it may not appeal to you, in which case you will decline it with your usual suavity, and I shall maintain my usual fine calm and sweetness of disposition. Yours sincerely, Arnold Bennett.[11]

He later agreed with her that theatre managers might be more impressed if she presented her plays under a pseudonym but added, 'you had better let me invent it for you. The ablest women are apt to give the show away at once when they choose a masculine pseudonym.' In fact, Lucy had already used a man's name – John Inglis – when she published her novel *George Wendern Gave a Party*.

After the runaway success of *Mrs Keith's Crime*, Lucy was able to indulge her love of travel – 'frisking' as she called it – on the Continent. In 1886, she and ten-year-old Ethel were with Lady Welby in Murren, and later the same year in Clarens on Lake Geneva, which she revisited several times, and where Ethel and Margaret continued their French studies. She used this continental setting for some of her short stories. Almost every year until her death she spent time in Europe. During these years she was reading for Macmillan and her own stories were appearing in serial form in *Temple Bar* and other magazines. In 1890, the year in which her husband's work *Seeing and Thinking* was published, she brought out her very successful epistolary novel: *Love Letters of a Worldly Woman*, which is discussed more fully later.

Edmund Gosse, who had known and respected William Clifford, regularly visited Colville Road and kept up his friendship with Lucy. In his book about Swinburne he wrote:

Professor W.K. Clifford early insisted on the intellectual importance of Swinburne's idealism, giving his lyrics a prominence which philosophers habitually begrudge to poets.

His letters to Lucy are interesting and outspoken. He encouraged her in her writing and always found something complimentary to say about her books. He rebuffed her quite facetiously when she tried to get him to meet a friend of hers: 'I am old. I am ugly. I am cross. I am no use to strangers. I have not long to live. Why should my last hours be tormented by an Esthonian lady. How would you like it if I asked you to go and call on a Latvian gentleman?' He was, with Lucy, one of the committee who set in motion the appeal to commemorate Henry James's seventieth birthday.

In 1899, Lucy moved further West to her final home at 7, Chilworth Street, just a matter of yards from Paddington Station. From the beginning of the nineties, her daughter Ethel had begun to be recognised as a poet of some promise. She held the Cambridge Scholarship at Queen's College in Harley Street, and her early poems had appeared when she was just sixteen in the 1891 *English Illustrated Magazine,* along with three short stories by her mother. This was one of many times when Lucy and Ethel's work appeared in the same publication.

In the spring of 1921, Queen Victoria's granddaughter, Princess Marie Louise, asked Sir Edwin Lutyens to design a doll's house for Queen Mary. The aim was that it should be something which would enable 'future generations to see how a King and Queen of England lived in the twentieth century, and what authors, artists, and craftsmen of note there were during their reign'. The astonishingly detailed and accurate miniaturisation of many of the artefacts from Buckingham Palace was carried out by the top craftsmen and women of the times. Of course there was a library, and the shelves were filled with miniature books. Every one of the well-known writers of the day who was asked to contribute a volume was delighted to accept – except George Bernard Shaw, who sent a brusque letter of refusal. Rudyard Kipling rewrote *If* and illustrated it specially as his contribution to the library. Lucy was among the chosen and she prepared extracts from *A Modern Correspondence, Letters of a Worldly Woman,* and *On the Wane.* Either Margaret or Ethel, whose handwritings were much more precise than Lucy's, completed the script of the tiny volume for her. However, she was even more proud that twenty-one of Ethel's poems, some of which were specially composed for the royal edition, were included in the Doll's House Library which can still be seen at Windsor Castle.

Both Lucy and Ethel contributed to the 1914 *King Albert's Book* – along with two hundred and thirty-two others. [13] The introduction stated: 'Never before, perhaps, have so many illustrious names been inscribed within the covers of a single volume.' Many, many thousands of copies were published with different editions and translations, in France, Russia, Italy and America. Very few of the contributors managed to avoid the sickly flavour of righteousness. But Arnold Bennett contributed a charming reminiscence of his first visit to 'the Continent', twenty years earlier. Two of the most interesting pages were written by Lord Reading, the Lord Chief Justice, and by another dear friend of Lucy Clifford's, the Right Hon. Augustine Birrell. Both entries make worthwhile reading today. The setting suited Rudyard Kipling's style as he savagely condemns the Germans with his poem *Outlaws*. Winston Churchill looks to the future, Frederic Harrison to the past, and the Aga Khan pledges the profound sympathy of the Moslems of the British Empire. The book is full of extraordinary contrasts – gems and horrors rub shoulders. Lucy's contribution is a rather cloying tribute as she addresses His Majesty King Albert, bowing her head to 'your divine example' and 'your crucified country'. Her daughter, Lady Dilke, has her poem *By the Lake* there too. It had been set to music by Liza Lehmann and was 'To be sung by Madame Clara Butt'. The song is sentimental but quite charming. Here is the first of the two verses:

> My son my little son, we two will rest
> Beside the water red in sunset light,
> And watch the evening fade, the lake grow grey,
> Until the moon's enchantment fills the night.
> Who knows what sombre fate stands near us now,
> What rich-robed destiny no eye can scan,
> What scented-sandalled Love?
> Ah! – Son of mine, play not with love when you are grown a man.
> Feign nothing and love greatly when you love.
> And having chosen, till the end be true.
> So shall one woman out of all the world
> Keep faith in man by keeping faith in you.

Ethel Clifford was a well-known and respected poetess. Her two books, *Lady Love's Journey* and *Songs of Dreams,* had been published by John Lane in 1905 before she married. Her poems and some short stories had been published in *The Westminster Review, English Review* and *Pall Mall Gazette*. Sadly, she gave up her writing after her marriage. Lucy tells a nice story of a chat with Tennyson when he asked her "Tell me did your child get any money for those verses of hers?" "Oh yes," I answered, "Macmillan sent her a guinea." He gave a delicious rumbling laugh, "Oh that is more than I got the first time!" He went on, "I used to write verses when I was a boy and my grandfather used to laugh at them and I would throw them away. Then one day our grandmother died and grandfather said, "Now write me some verses about your grandmother's death." I did and he read them and said "Why, they are quite good, here's a half sovereign for you. It's the first money you've ever earned and take my word for it it will be the last!" '[14] The Dilke family, with their country estates and fine London properties, always rather looked down on Lucy, and there was a distancing between Ethel and her mother after her marriage. Lucy's literary successes and unconventional circle of friends cut no ice in the Dilke's upper-class society world, and of course she was far from wealthy. Ethel's husband, Fisher Wentworth Dilke, was the son of a marriage between both sides of the notorious 1886 Dilke sex and divorce scandal. His father was Sir Charles Dilke's brother, Ashton, and his mother was Maye, the sister of Virginia Crawford, whose accusations of adultery had brought Dilke down. When Sir Charles's only son died, the title came to Fisher Wentworth, and thus Lucy's daughter became Lady Dilke. They had three sons but Ethel is not remembered as an affectionate mother. Her second son Christopher wrote: 'I didn't know my mother . . . . I knew my Grandmother [Lucy Clifford] much better and I really loved her. She was always surrounded by cats and wrapped in shawls and was tremendously kind.'[15] Ethel died in 1959 when she was eighty-three. Lord Kennet wrote her obituary for *The Times*. He commented that, though she was lost to poetry, she would be remembered as a 'gardener of genius' for the innovative horticultural developments on the Dilke estates at Lepe, Hampshire, and for her collection of miniature plants in the conservatory at their London home. Margaret, who, although she had poor health, took up nursing during the war and had appointments in London hospitals, mainly lived with her mother. Lucy was always concerned about her, and financed visits to Health Spas in Europe and clinics in London during her periods of ill health. She had always seemed to be overshadowed by her elegant, statuesque sister. The lovely drawing of Ethel by Edward Burne-Jones, which Lucy had reproduced as a greeting card, gives an idea of her striking beauty. Margaret helped her mother with her entertaining and did secretarial work for her. One of Lucy's oldest and closest friends was Sir Sidney Colvin, Keeper of the Department of Prints and Drawings at the British Museum. He was very fond of Lucy, praised her books, and wrote many kind and affectionate letters to her. He became fond of Margaret too, and even managed to make her nickname, Turkey, more dashing by calling her 'the Turk'.[16] After his wife died in 1924, Margaret visited him often, sorted out his papers and helped him through some very dark days.

Lucy Clifford had had an earlier connection with the Dilke family. She was for many years one of the women contributors to the 'gossip column' of *The Athenaeum*. This weekly, literary review journal, which eventually evolved into *The New Statesman*, had, from 1830-1846, been edited and owned by Charles Wentworth Dilke. As Marysa Demoor pointed out in her article *The Women of the Athenaeum 1890-1910*, when his grandson Sir Charles Dilke became its proprietor in 1869, the journal established itself as keen to accept contributions from women, particularly from 1885 to 1904 when Dilke, after his political career was in ruins, was married to Emilia Pattison who was a powerfully outspoken feminist.

Lucy had cultivated a friendship with a lifelong friend of Sir Charles Dilke who had remained constant to him through the terrible time of his court case and his resulting loss of power and position. This was the MP, barrister, politician, and academic writer, Lord George Edmund Fitzmaurice. Charles Dilke and Edmund Fitzmaurice had known William Clifford at Cambridge. Fitzmaurice was the second son of the fourth Marquess of Landsdowne. He had excelled at Eton and become a scholar at Trinity College where he won various prizes and got a First in Classics. His Presidency of the Cambridge Union Debating Society marked him out for a brilliant career in politics. With his command of modern languages, he was a natural for the post of Under Secretary for Foreign Affairs, which he first took in 1882. His ability in delicate diplomacy saw him negotiating European re-organisation in the 1880s. Then he got the Duchy of Lancaster and, in 1908, a seat, with Dilke, in the Cabinet. At the age of sixty-two, he was struck down by arthritis which left him a virtual cripple. He abandoned London and settled into what he called his 'buried alive' existence at Leigh, near Bradford-on-Avon in Wiltshire. Nearly two hundred letters written to Lucy by Lord Edmund George Fitzmaurice are held in the Valehouse Collection. They record one side of a most unusual friendship. In 1917, Lucy wrote to Fitzmaurice on the anniversary of William's death, anguishing about what he might have achieved had he lived. Fitzmaurice agreed that it was indeed a 'sacred anniversary'. In one of his rare shows of personal feeling, he wrote that he 'hardly took up the paper now without reading of the death of one or other of my Cambridge friends. I would love to see the old place once more which I associate with them – with nobody more than yr. husband.' Lucy and Fitzmaurice corresponded with each other regularly over more than a decade. His letters bear the mark of a man of habit and restraint. They almost all cover exactly four sides of notepaper. Without exception they begin and end with the same formal words. She is always *Dear Mrs Clifford*, he always *Yrs. Sincerely, Fitzmaurice*. There is no biography of this quietly eminent man. In June 1906, *Spy* caricatured him for *Vanity Fair* with the rather cruel caption 'He does not under-estimate his own ability.' Certainly, from his letters, one realises that he had very firm views on how international politics should be conducted and did not suffer fools gladly. Being himself an accomplished linguist one is not surprised that he once wrote disparagingly: 'Balfour's French . . . from what I hear is weak: but that does not signify in America. It is more serious in the case of Asquith, Grey and Lloyd George when they go to Paris and have to trust to interpreters'. Lucy was a thoughtful friend to him. Busy and short of funds as she usually was, she still found time to pack up and regularly post to him her copy of *The Nineteenth Century Review*, and articles from it would often provide the starting point of part of his letters to her. They would exchange notes about what they had been reading and Lucy passed on social chatter about who was in London. Her letters must have livened up his life at Leigh. His letters on the other hand were quite often reflections of how depressed he was about political affairs in Europe and in Ireland. His poor health and susceptibility to the damp and to hot weather were often reported, but a few of his letters do show how pleased he was to have her friendship. In spite of his political eminence, through the correspondence he comes over as something of a highbrow lame-duck whom Lucy loyally befriended with letters until just before her death.

It is amazing how Lucy managed to find time to keep up with her friends and her work. Bundles of affectionate letters from Augustine Birrell, author and statesman, Sir Ray Lankester, distinguished zoologist, Sir Edward Burne-Jones, famous pre-Raphelite artist, and Frederic Harrison the author, all of whom were of the same age as Lucy, are evidence of the devotion she inspired. Sir J.M. Barrie, playwright and novelist; Eliza Lynn Linton, writer; Lord Haldane, philosopher and statesman; Percy Lubbock, biographer who admired her books . . . the list of the friends she corresponded with seems endless. The fact that she never lost

Pencil sketch of Lucy's daughter, Ethel, by Edward Burne Jones. Lucy had the picture reproduced as a greetings card to send to close friends.

her youthfulness of spirit and courage made her always an attractive companion. She also never lost her love of travel, and in 1920 at the age of seventy-four she set off alone for Burgos in Spain, which William had visited as young man. From there she wrote a long and beautifully evocative letter to her dear friend Sydney Cockerell who was then director of the Fitzwilliam Museum in Cambridge and who had lent her books about Spain. She made no complaint about the heat, or the difficulties of travel, or the lack of London comforts, but made tender observations of the beggar-children and the gipsies who had invaded the town and the old professor who was giving her Spanish lessons.[17] Like so many of her heroines she never lost her sense of wonder at the beauty of the world and kept seeking, like them, the secret of harmony with nature, with mountains and with mankind. The year before she died, she received a small legacy and set off again through Switzerland to Italy for a last 'frisk' on the shores of Lake Garda. Her latter years were sad in that she was constantly offering her work to publishers and producers and having it rejected. But, to the end, her interest in life and in young people did not fail her. She fell ill in March and died on 22 April, 1929. The cause of death was given as bronco-pneumonia and cerebral thrombosis. She had asked for her occupation to be entered on the death certificate simply as 'widow of William Kingdon Clifford'. Her property and personal estate amounted to just over £2,500. In all of her many obituaries, tribute was paid to her generous nature, but perhaps Henry James's comment in his diary about Lucy's 'great personal sweetness and frankness and niceness, and her heroic and arduous life, her admirable courage and labour and gaiety and independence through it all'[18] gives the truest personal picture of this remarkable woman.

It is not difficult to see why her books have not lasted. By mainly limiting herself to writing about personal relationships in a defined middle-class milieu, her themes were bound to seem dated as soon as the break-down of the old order took place. In spite of her husband's influence, her books took no account of the excitements of the scientific discoveries and educational opportunities of the age, and, although she strove for freedom for women, it was a non-militant striving for non-assertive freedom. However, though her settings may have been conventional she was never banal. Her high sense of romance, her nice appreciation of character and motive, her eagerly impulsive interest in life – these were her stock in trade. She loved life, was loyal to her friends, and followed to the end her husband's philosophy – she did what she did best as best it could be done.

Notes

1. ALS, Cheshire Record Office, Delves-Broughton Collection, Box M.
2. ALS (one of 15), Valehouse Collection.
3. Irene Cooper Lewis (ed.) *Vernon Lee's Letters*. Privately printed, 1937.
4. ibid.
5. ibid.
6. ibid.
7. There are 48 letters from Lucy to Karl Pearson held in the Library archives at University College, London, (Pearson 661).

8. ALS (one of seven), Valehouse Collection.
9. Joan Hardwick, '*An Immodest Violet' The Life of Violet Hunt*, Deutsch, 1990, p.120.
10. ALS (one of seven), Valehouse Collection.
11. ibid.
12. ALS (one of six), Valehouse Vollection.
13. *King Albert's Book: A Tribute to the Belgian King and People,* published by *The Daily Telegraph,* 1914.
14. Notes for a talk to young writers found in Lucy Clifford's papers.
15. ALS, Valehouse Collection.
16. ALS (one of twenty), Valehouse Collection.
17. Viola Meynell (ed.) *The Best of Friends, Further letters to Sydney Carlyle Cockerell,* Rupert Hart-Davis, 1856. p. 22. Meynell describes Lucy as 'A charming and highly cultivated woman who had many friends in Cambridge with whom she stayed from time to time'.
18. L. Edel and Lyall H. Powers (eds.), *The Complete Notebooks of Henry James,* Oxford University Press, 1987, p. 350.

Part Three
Heritage: Literary and Scientific

Lucy Clifford's Books and Plays

Clement Shorter in his 1897 book on Victorian Literature comments:

> I might easily devote many pages to the living women novelists who have impressed themselves upon the era; but that scarcely comes within the scope of this little book. There are, to name but a few, Mrs Lynn Linton, Mrs Humphry Ward, Ouida, Miss Braddon, Miss Marie Corelli, Miss Olive Schreiner, Miss Rhoda Broughton, Edna Lyall, Lucas Malet, Miss Charlotte Yonge, Miss Adeline Sergeant, Mrs Macquoid, Mrs Alexander, Mrs W.K. Clifford – names which recall to thousands of readers many familiar books and some of the happiest hours they have ever spent.

Most of the writers on that list, and many more of that age, have suffered the same fate: their books were popular and sold in their thousands, and yet it is quite difficult to find a copy of one of them today. The style in which they are written seems dated, and the content often without relevance to the modern reader. However, there have been some notable survivors and revivals – especially through the Virago Modern Classics series. Short stories by Lucy Clifford have been included in several modern anthologies.[1] But, in order to understand their impact and influence, we need to look at the way in which her books were received and reviewed at the time of their publication.

In her writings and professional life after William's death, Lucy was always known as Mrs W.K. Clifford. She always wore black and said that she felt herself still married for the fifty years that she was without her husband. Before her marriage, she had begun her career as a writer for newspapers and magazines, and her novel-writing, under her maiden-name Lucy Lane, had already taken off with serialised stories in *The Quiver*. Her stories appeared in each edition between 1871 and 1877. Her first effort: *The Troubles of Chatty and Mollie* had appeared Volume VI, 1871. The ten instalments of *About Nellie* came in Volume VII, 1872. The same volume also contained the twenty instalments of *The Dingy House at Kensington* which Lucy later rewrote and extended for its appearance as her first published book.[2] *The Saturday Review* gave it detailed attention and reviewed it as 'somewhat scrappy and improbable' but 'giving promise of better things to come'. It was published anonymously and any first-time author would have been well pleased with the reviewer's final paragraph:

In conclusion, we cannot too highly praise the healthy tone of the whole story. Although the drawing of the characters is both thorough and delicate, it is commendably free from that super-subtle analysis of hidden motives that makes many modern novels so indescribably tedious. The style is good throughout; simple, bright and unaffected, and happily devoid of any pretence at 'word-painting'. It sounds incredible, but we do not recollect a single description of a sunset.

Praise can no further go.

This was the first of the many books and plays that established her as a popular, but never first-rank, writer. She was successful enough to earn a mainly steady income, but never enough to be quite free of financial worries. She admitted to Virginia Woolf in 1923 that she had always had to work, had been in debt, and had 'sold the furniture often, but never borrowed'.[3]

No single classifying label can be attached to Mrs Clifford's writings. The early reviewer who anticipated 'better things to come' could never have foreseen the extraordinary mixture of themes, predicaments, pressures and plots which she would tackle in her writings, or have anticipated the techniques she would employ to illustrate them. She never used her books as a platform for her unembarrassed agnosticism, but William's conviction that there was no need for other than human strengths to meet human needs comes clearly through in all she wrote. She avoided sentimentality. The freedom she wanted for her female characters was not militant and political but personal. She, and the women she portrayed in her books, wanted the right to exist, to feel, to think and express their thoughts, and, if they wanted, to marry without having to surrender such rights. They wanted, like Isabel Archer in James's *Portrait of a Lady*, to be free to 'choose', and the fact that some of her characters have similarities with some of those portrayed by James has been noted by other critics.[4] Lucy Clifford wrote to earn her living and needed to please the public, but the introduction of these views into her writing set her apart from many women writers of Victorian/Edwardian fiction.

Her first children's book after William's death, *Children Busy, Children Glad. Children Naughty, Children Sad*, was published in 1881. It was a collection of verses most of which have a moralistic strain to them and will seem dated to modern readers, but they sold well – 35,000 copies – and were translated into German. The book was first published simply under the initials L.C. In November 1881 Lucy received a letter from Mark Pattison, Rector of Lincoln College, Oxford. In it he warned that:

People here who profess to know things are giving your verses to Lewis Caroll (sic) and that is in spite of his own disclaimer. The initials, you see, are the same. Will you not do something to claim your own property?[5]

This was a valuable misattribution and Lucy did not hurry to set the record straight. She certainly used the fact that her book had been linked with Lewis Carroll's name to impress Scribners when they were publishing her books in America. All of Lucy's early books and short stories were published anonymously.

However, she made an exception of *Anyhow Stories: Moral and Otherwise,* and, for the very first time, signed herself as Mrs W.K. Clifford.[6] She did this because she wanted it known that, in this book, she was carrying through ideas she had discussed with William. She made some changes to the second edition.[7] She dedicated it, as if William were still alive, 'To You and the Children'. It is enhanced by the quaintly nostalgic illustrations by Dorothy Tennant and the Hon. John Collier. This book is a curious mixture. Many of the stories certainly bear the mark of William Clifford's philosophy and the influence of Mazzini upon him. They suggest that life is a workshop where all must try to make something good and beautiful according to their abilities. The deeply psychological sophistication of one or two of the tales and verses been noted in recent critical appraisal.[8] The first edition contains ten stories and three poems and is prefaced by the explanation that the stories are 'not meant only for very little children, but for any who may care to read them'. In fact, the book would have better been prefaced with a warning: 'Read at your peril!' for among its mainly harmless contents is one story that even to today's readers, hardened by television, film and newspaper presentation of frightening events, would still find hard to take. In 1996 a suggestion, carried on the World Wide Web, that Lucy Clifford's stories would be of interest to anyone studying Victorian children's books, set off alarm bells for at least two subscribers who sent back warnings that *The New Mother* was a horrid story that could give nightmares to the susceptible. One subscriber 'would not read it again for any money'. Another agreed with the warning and suggested that it was suitable only for the really intrepid reader.[9] Others have been seriously affected. Harvey Darton, the publisher, wrote 'Getting on for fifty years after I first met her (The New Mother), I still cannot rid my mind of that fearful creation.'[10] Robert Lee Wolff recorded it as a 'terrifying' story.[11] More recently Lord Annan wrote privately of how this horrifying story, read to him in childhood, had remained with him into his adult life. It has also been described as: 'Possibly the most extreme example of pointless cruelty in a century that abounded in terrifying stories for the young'.[12] How did it come to be written? It is inconceivable that William Clifford himself suggested it. His fantasies for children took the form of nonsense situations and ludicrously funny inventions. Nothing in his philosophy allows for the cruel irreversible punishment for naughtiness which *The New Mother* contains. The story begins in a modest village home. The father is away but the two little girls, their mother and the baby are contented and happy. The little girls on their way to the village meet a strange girl who tantalises them with a story of a tiny dancing man and woman who can only be seen by truly naughty children. They become obsessed with curiosity and deliberately commit naughtier and naughtier acts in order to qualify for a chance to see the tiny people. Their mother, in desperation, threatens that if they continue to be so naughty she will have to leave them and be replaced by a new mother. She describes this replacement mother as having glass eyes and a wooden tail. The inevitable happens. The children behave even more shockingly and their mother leaves them. The final twist is that the children still do not see the tiny dancing people and the reader comes to realise that the girls have had an encounter with an evil presence with the sole purpose of compelling them to destroy themselves. They ask the stranger if their mother will come back:

'Never,' sang the girl; 'she'll never come – never come again. She is sailing away to the sea; she will meet your father soon; they will sail – sail – sail to the countries far away – very far away –.' When they heard this, the children gave a cry, but could say no more, for their hearts seemed to be breaking.

The horror of the ending of the story is that the reader or listener to the tale is left totally without hope. The New Mother is approaching the cottage. 'She had a black satin poke bonnet with a frill around the edge, and a long bony arm carrying a black leather bag. From beneath her bonnet flashed a strange bright light'. The two little girls inside the cottage are terrified:

So together they stood with their little backs against the door. Then they heard her say to herself – 'I must break open the door with my tail.' . . . For one terrible moment all was still, but in it the children could almost hear her lift her terrible tail, and then, with a fearful blow, the little painted door was cracked and splintered. With a shriek the children fled . . . into the forest beyond.

The story ends:

They are there still, my children. For long months have they been there, with only bracken for their pillows and brown dead leaves to cover them . . . . Now and then, when the darkness has fallen and the night is still, they creep up near to the home in which they were once so happy, and, with beating hearts, they watch and listen; sometimes a blinding flash comes through the window, and they know that it is the light from the new mother's glass eyes, or they hear a strange muffled noise, and they know that it is the sound of her wooden tail as she drags it along the floor.

The children, in an almost Faustian way, have succumbed to temptation and paid a terrible price. Even accepting the familiar Victorian themes of children being brought to repent by punishment or by horrible threats, it is not easy to understand how Lucy came to write, to read aloud, or to publish, such a horrifying tale. She then compounded the horror, by giving the pet names of her own daughters – 'Blue Eyes' and 'The Turkey' – to the unfortunate pair of girls. It was published when Ethel and Margaret were eight and seven respectively. Even delivered in the cosy comfort of a mother's arms, *The New Mother* could never be anything but dreadful. That Lucy truly loved and cared for her 'chicks' is transparently clear from her letters. But she was alone, she had embraced her husband's atheism and denied herself the comfort of religious faith and thus flown in the face of conventional thinking. In this story she faces out her own worst nightmare – that of losing charge of her children – and, through the example of *The New Mother,* she is attempting to bind her children closer to her. Another of the *Anyhow Stories* that continues to attract attention today is *Wooden Tony*. This was originally published in 1892 in Lucy's collection of stories for adult readers: *The Last Touches and*

Other Stories. It too, is an extraordinary tale and has recently been republished.[13]. The story is set high in the mountains of Switzerland where the little boy Tony is considered a 'wooden-head' because he is a dreamer and lover of nature who wants to remain tiny and so avoid the problems of growing up. He sings songs that nobody has ever heard before and becomes more and more withdrawn. His father carves wooden figures to sell to tourists and Tony comes to wish he was like them – small, protected, and kept safe in a drawer. The story ends with him being taken to Geneva by a dealer and gradually shrinking, losing the power of movement, and eventually joining his father's carved wooden figures in a cuckoo clock where he is jerked into view as the hours strike and his song is played. His parents see him in a shop window where he is condemned to this routine: 'Hour after hour, day after day, week after week, month after month, in light and dark, in heat and cold.' Again the story ends with a desperate note:

> 'It is Tony,' cries his mother pointing to the clock.... 'It is his song – he is wooden.' While she spoke the song ceased, the figures were jerked into darkness, and the doors closed; before the man and the woman lay the long road and weary miles that led back to the village and the mountain.

Modern critics have given *The New Mother* and *Wooden Tony* strong underlying psychological interpretations. An American psychoanalyst has seen *Wooden Tony* as a study in childhood autism. In a twenty-two-page analysis of the story he conjectures: 'The poetic insights of Victorian literature such as are found in the tales of Collodi (who wrote *Pinocchio* in 1881) and Clifford, can be seen as forerunners of psychoanalytical thinking.'[14] It seems most likely that Lucy got the idea and setting for the tale from one of her visits to Switzerland, and she has an Englishwoman staying at an Alpine hotel near to Tony's cottage in the story. *Wooden Tony* was written eight years before Freud published *The Interpretation of Dreams* and it seems unlikely that Lucy consciously intended to chart the child's decline through depersonalisation and alienation to eventual dehumanisation; and yet it is that process that is so graphically observed and depicted in the story.

Lucy Clifford's work has attracted some attention in recent years. *The New Mother* is included in the 1993 *Oxford Book of Modern Fairy Tales* edited by Alison Lurie. Lurie points to parallels between this story and *The Turn of the Screw*, written over a decade later by Henry James, and asks 'It would be interesting to know whether James, when he wrote his famous ghost story, remembered his friend Lucy Clifford's strange and haunting tale for children.'[15] Even though the germ of the idea for James's chillingly evil tale came from a conversation with E.W. Benson when he was Archbishop of Canterbury, it is highly probable that he and Lucy chatted together about *The New Mother* and its disturbingly ambiguous themes. The story has also recently been included in *The New York Review of Science Fiction*.[16] More recently, in 1990, one critic has suggested that Lucy's story is a reactionary late-Victorian tale portraying the middle-class family at risk from the contagion of disorder bred among the lower classes, which only strict adherence to the mores of Victorian ideology can avert.[17]

But these possible interpretations must be considered again in the light of Lucy's newly revealed correspondence. We find her, in 1910, writing to tell her publisher that she had rewritten *The New Mother* as a 'musical comedy for children', introducing some more characters, and – impossible though it may seem – a happy ending. She also suggests that *Wooden Tony* could be set to music. It is perfectly clear that, even if Lucy wrote *Wooden Tony* and *The New Mother* with the intention of demonstrating deep moral and psychological significances, she had no great loyalty to those themes, and was prepared, even keen, some thirty years later, to cannibalise them and produce totally different ones. It was during these later years that Lucy's attempts to interest publishers were becoming desperate, and her diaries reveal string upon string of rejected manuscripts and scenarios.

Lucy Clifford's ability to step aside from conventional themes and techniques was startlingly demonstrated in 1885 when everything she had written previously was overshadowed by the sensation caused by the appearance of her first non-serial book. *Mrs Keith's Crime: a record* was the talk of the town and remains the title by which she is most remembered today. Its subject, infanticide, was deeply shocking but it was not just a case of an opportunistic grab at the headlines. She had first shown it to Lord Morley who had been a close friend since the time that he and William had been members of the Savile Club. He thought the fact that it was written in the present tense was 'a literary offence'. He also thought the end 'revolting'.[18] Macmillan and Blackwood had refused to publish it.

Eventually Richard Bentley produced it ion two volumes. For a later edition the Hon. John Collier provided a charming 'imaginary portrait' of Mrs Keith. From the earliest pages it is clear that Mrs Keith is bound to be overtaken by disaster. She is artistic and unworldly, a lover of nature, trees and the countryside and, like Lucy, she inspires friendship. Her husband is drowned and she, like Lucy, is left with two small children. Signs of her own disease are appearing as she plans to travel to Spain to seek a warmer climate for her ailing daughter, but her son falls ill and dies first. The wealthy Jew, who befriends and financially assists Mrs Keith after the death of her healthy son, clearly cares for her, although she does not see this. Several literary critics congratulated Lucy on her life-like portrayal of Mr Cohen, and for inventing the 'only gentleman Jew in literature'. The handsome young doctor, whom she is fortunate enough to consult in Malaga, falls in love with her. But Mrs Keith at first does not recognise his feelings and later nobly conceals her own feeling so that he will return to a prestigious appointment in England. She becomes obsessed with her daughter and realises it, 'For love has no bands that bind so close as those that fear and sorrow weave.' she gradually separates herself from those who might help her. When her own poor health fails and it is pronounced certain that she will die first, she decides to kill, with chloroform, her daughter rather than leave her to die in the care of strangers. Even today, this bitter choice is agonising to contemplate.

In the year 1885, however, even to discuss such a possibility was quite beyond the pale. This was the age when, as Lucy graphically portrays early in the story of Mrs Keith, a divorced woman was a guilty woman. Even to behave in a friendly, sympathetic manner towards a divorcee was contrary to conventional

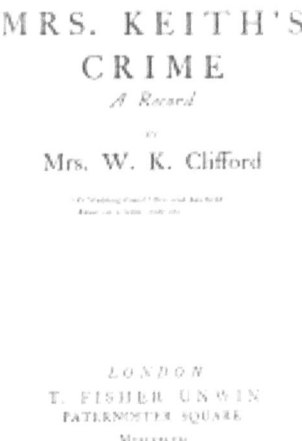

John Collier's frontispiece in the Fisher Unwin edition of Mrs. Keith's Crime. *Collier illustrated several of Lucy's novels and serialised stories.*

behaviour for many people, and Lucy was to return to this cruel injustice many times in her writings. How much more shocking it was – even in a novel – to attempt to paint a sympathetic background to the most repugnant of all crimes, infanticide. By writing the book in the first person and using the present tense she rendered Mrs Keith's dilemma even more compellingly distressing. She prefaced the book with these words:

> It would be useless to try to account for the manner in which this history came to be written down. It is obvious that Mrs Keith's hand could not have written it, nor could her voice have given it utterance, and there was none by her in that hour when her love gave her terrible strength, and left her to brave eternity. It seemed almost as if, as she passed along, the air itself bore witness and the wind swept into the heart of one who understood all the unspoken thoughts of that passionate life.

The decision that she should publish anonymously was both prudent and astute. There was a furore, there was a lot of speculation, and the book ran to four editions before Lucy was revealed as the author. Robert Browning thought it 'splendid' and said he wished he had written it himself. Thomas Hardy later wrote to Lucy recalling how Lord Houghton had enthused about it at a dinner party at the home of Sir Francis Henry Jeune.[20] It was discussed everywhere, and

Lucy was firmly established as a novelist of strength. The fact that Mrs Keith had no religious faith was shocking to many. Even in her deepest anguish she never prayed or called for divine forgiveness for the crime she intended to commit, and, after committing it, calls only for mercy from her dead husband. *The Church Times* reviewed it as, 'original, powerful and *dangerous*' and concluded that:

> so great is the writer's skill that our horror of the mother's crime is quenched by our deep pity for the unhappy woman to whom, being a Pagan, the glory of submission to the Divine Will is unknown.[21]

Those last few words might well have been spoken by critics of William Clifford. Other critics wrote that it was 'a daring piece of work', that it was 'little short of genius' and 'full of distinction'. The passage which links Mrs Keith most firmly to Lucy's own experience and suffering comes early in the story soon after the death of her little boy from Scarlet Fever, caught from his nurse. Mrs Keith remarks to the childless, unloved woman at whose home her son died: 'It is a terrible thing to be born with a great capacity for happiness and to feel that one has hardly had one's share'. The woman then asks:

> 'Did your husband love you?' 'Oh yes, yes' I answer, 'With all his heart.' 'As you did him?' 'As I did him,' I answer. 'And you had your children while he was here?' 'Yes, we have laughed for joy as we watched them.' 'Then be satisfied,' she says gently; 'You have had your share. If you had possessed all you have had for just an hour, you would have had your share. There are many women with as great a capacity as yours for happiness, who are hungry for it all their lives, yet are never satisfied for one single hour.'

The last seven pages of the book are presented as a 'stream of consciousness' far ahead of its time, but admirably suited to expressing the emotional crisis of a dying woman preparing to take her child's life:

> I pull the shawl closer and closer round me . . . it is the one I am going to wear while I kill Molly . . . for night has come, this fearful night, on which no morning must ever dawn for us . . . . Oh! Am I mad? But no – no; I am sane enough and know too well – too well – . . . . Molly . . . if I could but kiss the life out of you . . . . I must go, I must make haste, for the strange pain is coming and I am growing cold already . . . . So many die with the night, and the darkness has come to sweep me away . . . . So – so – die in mother's arms, my own – mother who loves you so . . . not those cold hearts . . . mother, who can do this – even this – this for love of you. I wait and gasp and listen . . . and raise my head once more and give one wild look round the room . . . it is empty . . . and still and desolate. Someone is here . . . . You! . . . Oh, speak . . . but you cannot hear . . . but no, the gulf yawns wider and wider – There were no hands to take her when mine fell . . . oh speak . . . have

mercy! – mercy! – mercy!. . .Only a cold wind that wandered round the empty room and swept out towards the empty sea . . . alone . . . the . . . the darkness is coming and something sweeps me away . . . . Jack is singing . . . . Jack, my sonnie, where is Molly . . . . I stretch out my hand . . . . Molly – baby Molly . . . gone . . . gone – Molly.

Sir Walter Besant, the novelist and historian, read *Mrs Keith's Crime* at one sitting and wrote to Lucy:

> I have to thank you most profoundly for your book over which I spent the whole of yesterday evening. You have discovered an entirely new situation in fiction and one of surprising strength. Your book shall live and not die. An old hack like myself feels ashamed at the freshness and intensity of the work. I declare that I have not been moved by anything so much since I was a boy and first read the great works. You can afford to laugh at the critics in spite of the present tense which is however so detestable in the hands of the imitators of Rhoda Broughton.[22]

Richard Bentley paid her £75 for the first year's sole rights, and a further £30 for the first 750 copies. Lucy retained the rights of translation, the Tauchnitz rights and the American rights. For the sixth edition Lucy attempted to answer some of the criticisms levelled at the book. She defends her use of the present tense as using the best tool for telling the story. She explained that she had felt herself almost possessed by an unseen woman. She felt herself to be a 'breathless spectator' and was compelled to finish writing it as if 'Mrs Keith' had: 'used my pen to tell the desperation and the anguish that drove her to that last act'. It became a thing 'for which I was not wholly responsible'. It was a 'defence for a dead woman, unconsciously uttered by her to any whom it may concern'. The fact that the book was so successful had not removed the pain of the critical reviews, and Lucy took six paragraphs to justify her presentation of such a tragic story. In 1922 Lucy Clifford asked her old friend Thomas Hardy if he would write a preface for a new edition. His letter in reply included these words:

> Alas, you are asking me to do what is mentally – and almost physically – impossible to a person of eighty-two . . . it is not a question of willingness, but of practicability, and not one of the many requests for the same kind of thing that I get have I declined with such regret.[23]

A new and revised edition was eventually published in 1925 with no new preface. Lucy did not attempt to repeat this highly individualistic writing technique but took another innovative approach to achieve her next big success. In 1880 she had begun her friendship with James Russell Lowell, which is described more fully elsewhere in the book. Much of their friendship was conducted through correspondence and her letters to him impressed him so much that he wrote:

I am to write a life of Hawthorne if I can. But what care you? This is more to the purpose, that I still think that you could write a successful novel in the form of letters – all the more now – that I wish I could see you, and that I am yours faithfully, J.R.L.[24]

The result of that suggestion was her publication, in 1890, of *Love Letters of a Worldly Woman*. This was serialised anonymously in *Temple Bar* and attracted attention from the start when authorship was first attributed to Oscar Wilde. Edward Arnold brought out the first edition, Bentley did two later editions and it came out in Constable's cheap editions during the 1914–18 war. The fact that it was not copyrighted in America, where it did very well, resulted in twelve pirated American editions. Some twenty or so years after its publication, when Lucy visited New York, she was lionised as its author. The book takes the epistolary form of exchanges between lovers and recounts the experiences of:

> three women who loved the world: not meaning (at least two of them) the pomps and vanities but the round world itself and the people who belong to it. All had the bandage lifted from their eyes and, as they became wise, proved how sad a thing is wisdom. The first tried to comfort herself with dreams; and waits hoping that they will find their way into the waking hours. The second plays an eager restless game, staking all her happiness on it; and perhaps gained most when she had lost it. The third looked up at sorrow and, seeing a little way beyond, set out on a journey; but she does not know yet where it will end. And the moral is – but who cares for morals nowadays? Let us leave them to the preacher.

The publication in the States of an 1894 'marked edition' promoted even more interest. It carried a Prologue:

> A certain society woman, whose exterior life seemed one of unalloyed contentment . . . one day saw this little volume and took it silently into her confidence. Page by page she marked the phrases which seemed to accentuate the emptiness of her own life – the yearnings of her impassioned soul – the very weariness of existence itself, then – like a mirror which disappoints us with our own imperfections, she laid it aside until she could again bear the company of her own reflection.

Readers were intrigued by this additional anonymous commentary – some attributed it to Edith Wharton – and pirated editions were said to have been sold on street corners in New York. The letters themselves chart different stages in relationships between lovers. In the first, *A Modern Correspondence,* the six-letter exchange is between a man and his woman friend, who has realised, after a holiday flirtation, that he is dull, conventional and unintelligent. Her letters display her idealism while his betray his shortcoming. They have totally different ideas of passion. His is 'wickedness', hers is the passion of Joan of Arc as she rode to Rheims, or of Napoleon striding before his army. She tells him; 'read St

Agnes Eve, – Tennyson not Keats, and you will understand'. She can't bear the idea of a life so dull that she 'could see the whiteness of her own tombstone at the end of it'. He says, 'I do reverence you for your goodness.' she retorts: 'How deadly dull to herself must a good woman be . . . . I do not want your reverence – it goes to passion's funeral.' Her ideal man is a free-spirited philosopher – like William Clifford, although, of course, he is not mentioned. She could 'tramp the mountain tops with a beggar who was a poet, a mechanic who was a genius, a dreamer who talked of a waking time to come . . . a man who is part of the great machinery that models the future ages'. The girl in the letters is expressing Lucy's ideals, and she vigorously and eloquently rejects the man who doesn't believe in Darwin and thinks women should not read Zola! J.R. Lowell wrote to Lucy soon after the book came out, and his last sentence in this unpublished letter showed how much this Lucy-like heroine appealed to him:

> I read your 'Correspondence' with great interest, as you may suppose, and was glad that I had a share in the impulse that pushed you into writing it. It is full of life and go. And those two are the springs of pleasure that never run dry. I have no quarrel with your heroine, who maintains her thesis as eloquently as Heloise could have done. I think if I had been the wooer I would have had her or died.[25]

This first section of the book was translated and three editions came out in France. The main section of the book, from which comes the title, is next. It tracks Madge Brooke, another free-spirited woman, through the break-up of her love-hate relationship with a fickle man who marries someone else and then, when she dies, tries to re-kindle his affair with Madge. She disdains him and eventually decides to settle for the companionship and affection of a loveless marriage to an elderly politician. An interesting comment about *Love Letters of A Worldly Woman* was made in *The Review of Reviews* 1892:

> It is one of the most delicate, most original, and most noticeable books of the season . . . . We would particularly recommend it to the aspiring male novelist. It will help him to understand women as nothing else will.

Lucy changed tack and theme yet again for her next success and surprised her readers with *Aunt Anne*. In February 1892, she wrote to her publisher Richard Bentley:

> It has no heroine. It is about an old lady of 68 who falls in love with, and marries, a young man of seven and twenty. Of course he breaks her heart and she dies in the end. She is rather a tiresome old thing though she is lovable and pathetic in her way. It would have to be very anonymous indeed for 'Aunt Anne' is alive and her love-making is going apace.

The meaning of the last sentence is obscure. However, a clue might be found in the fact that Baroness Angela Georgina Burdett-Coutts, the wealthy philanthropist and close friend of the Pollocks, had, at the age of sixty-seven, caused quite a stir by marrying her thirty year old secretary and perhaps Lucy felt that an embarrassing parallel might be drawn. In the end however, the story did come out under Lucy's name. It was serialised in *Temple Bar* and the book came out in July 1892. One thousand copies were first printed with 500 each for the first and second edition. Lucy squeezed a good contract out of Richard Bentley, getting £200 for the copyright and £55.10 for 1500 copies of the sixpenny edition. He ended his letter of agreement: 'From which you will perceive that the fair sex make better bargains with us than its masculine rivals.'[26] It was a great success and remains one of her best-known works. The book has no dedication. However, James Russell Lowell had died just the year before publication and she chose to place one of his verses in the frontispiece. It is a story of emotional blackmail. What begin as quite trivial demands made almost absent-mindedly by Aunt Anne on the young couple who are too nice to stand up to her, become serious enough to threaten their financial stability. The incorrigible Aunt Anne, who gets her own way in everything, is more than paid back when she discovers that the young man she marries is only after the inheritance he imagines will come to her. Arnold Bennett summed Aunt Anne up as a 'disreputable old charmer for whom he had a lasting regard'[27]. When his most famous book, *The Old Wives' Tale,* came out in 1908, he justified writing about older women in his preface, and said: 'I had always been a convinced admirer of Mrs W.K. Clifford's most precious novel, *Aunt Anne*; but I wanted to see in the story of an old woman many things that Mrs W.K. Clifford had omitted from *Aunt Anne*.' Aunt Anne is Mrs Baines, and Bennett curiously chooses Baines as the family name for the two wives in his novel. Lucy's book came out as a 'double-decker', was reprinted many times and translated into German. Lucy said that she had had countless letters about *Aunt Anne* from people saying that they recognised their own particular Aunt and asking how she came to know her. A similarity has been noted between Lucy's Mrs Baines and Henry James's Miss Tina in the 1889 *The Aspern Papers*.[28]

In 1892 Gladstone became Prime Minister for the fourth and last time. He read *Aunt Anne* and wrote some kindly comments about it to Sir James Knowles who passed the letter to Lucy. On the strength of that she sent Gladstone one of the first copies of her 1896 autobiographical novel, *A Flash of Summer.* She explained that, through the book, she was agreeing with Gladstone's views and pleading, not for easier divorce, but that marriage should not be 'a matter of convenience or of persuasion against a weaker will'. Gladstone read *A Flash of Summer* and wrote an encouraging letter to Lucy expanding his views. In Lucy's reply to him she suggests to him that he should write a book urging greater reverence towards the institution and ideals of marriage. She wrote: 'If a great man urged it, it might do a world of good. It would become a text-book.'[29] The Bishop of London, Mandell Creighton, after reading *Aunt Anne,* wrote to Lucy saying that 'My historical studies among Popes has shown me that what needs explanation is the impetuosity of old age . . . you have had something of this in your mind; the spring of hope is eternal, and it is hard to believe that all things are not possible.'[30] Lucy's manipulative yet manipulated, ageing heroine found her way into the hearts of a wide range of readers and *Aunt Anne* ran to five editions and eventually came out in Macmillan's sixpenny editions. It was considered by many to be her finest book. After reading it, Thomas Huxley gave Lucy a copy of his book *Essays Upon Some Controverted Questions*. He dedicated it 'from Pater. October 1892', and then wrote:

> What one can't
> T'other can.
> I do Science,
> You do *Aunt Anne*.[31]

During the 1890s and early 1900s, Lucy wrote dozens of short stories for magazines and journals. Over thirty of these she gathered together for her several editions of collected stories. In 1892 her short story *A Grey Romance* appeared.[32] The lonely and middle-aged man and women in the title story are attracted to each other and drift into a relationship. On the eve of marriage they each realise that they will be destroying for the other the very thing each treasures most – freedom from involvement. She, the romantic dreamer, flees to the continent. He, the constrained and conventional lawyer troubled by neuralgia, accidentally overdoses and dies. Neither knows of the other's actions. Three more of Lucy's stories were published together in 1901. They are prefaced by a line from Browning, 'A turn, and we stand at the heart of things.'[33] In each of them it is the proposal of marriage that brings the characters to reveal their true feelings or secrets. Many of Lucy's short stories centre on this predicament. Her story *Miss Williamson* is interesting in that it has the same theme as the story of Lydgate's first love in *Middlemarch* published thirty years earlier. Lydgate fell in love with Laure, a woman acquitted of murder who, as he proposes marriage, reveals that she was in fact guilty. In Lucy's shipboard romance Miss Williamson flees after her lover's proposal of marriage, leaving him with a newspaper account of her acquittal with her confession of her guilt scribbled on it. In another of her books, *Marie Zellinger,* Lucy attacks the conventional attitudes of London social life by allowing an intellectual, attractive and unconventional Austrian girl to put the cat among the stuffy pigeons with her bohemian behaviour.

In 1893, an interesting exchange of letters appeared in the *Athenaeum*. Lucy was very annoyed to find that, without her permission or knowledge, a semi-religious story of hers, published anonymously sixteen years earlier in *Quiver*, had been retitled and republished by Warne. It came out in book form as *Marie May or Changed Aims,* and rode on the back of her success with *Mrs Keith's Crime*. Lucy came out strongly on the attack with a letter headed, '*A Warning to Authors'*. She felt that it was sharp-practice to advertise this early and slight work as by the author of *Mrs Keith's Crime, Aunt Anne* and *Love Letters of a Worldly Woman*. Warne defended themselves equally strongly – they had after all bought the copyright of the original story. Lucy had the last word with a long rejoinder detailing her objection to being judged by a story written long before she had become a successful author. *Marie May* – a novel in which class and the differences between gentlefolk and 'trade' hamper the course of true love, would clearly disappoint those who had enjoyed Lucy's later novels. Lucy had formed, early in her career, a strong link with the publishers Scribner's Sons of

New York. She met and made friends with Mr Scribner and his wife when they were in London, and many of her stories appeared in Scribner's Magazine. The Scribners loved receiving her letters which were full of London literary gossip. Through them she made contact with film companies. Warner Bros made a film of *The Likeness of the Night,* but Lucy thought they had made it 'so grotesque that I was ashamed of seeing my name to it'. She received $1,500 for the film rights of her story: *Eve's Lover,* but it was never filmed. Scribner regularly sent her books to review and she tried them out on her famous friends and reported their reaction.[34]

Travelling was Lucy Clifford's passion and she drew on her experiences abroad to add colour to many of her novels. She admitted that when seeing friends off at the station she always had a 'longing to put myself in front of any great truck full of luggage, and stalk along with a martial air as if it belonged to me and I were going off on my travels. I love the change, and the people, and the scenery – above all the mountains! There is nothing like them in the world.'[35] She wrote travel articles for various newspapers and for *Quiver* magazine. One of her own favourites among her books was *A Wild Proxy* published in 1893. This was a light-hearted comedy of errors, whereby a runaway couple go rushing around on trains and boats visiting Paris, Marseilles, Nice and Genoa before returning to London to face the music. She put her foreign travel to good use again in one of her most popular later books. *Sir George's Objection* was set and written at her favourite Italian holiday haunt, Cannero, on Lake Maggiore. It was quite moving to find, in 1996, that the proprietor of the Cannero Hotel, where Lucy had stayed while writing the book, still treasures her great-grandfather's memories of the lovely English lady who used to sit in the shady garden at the back of the hotel to write her stories. Lucy evidently quite won their hearts and the family still treasures copies of her books, which she presented to them. Cannero, its streets and buildings, gardens and landscapes are instantly recognisable from Lucy's descriptions of them. The book is about skeletons in cupboards. Will the marriage of the beautiful, widowed Mrs Roberts's daughter and Sir George Kerriston's son be forbidden if the dark secret of her dead husband's criminal record is divulged? Mrs Robert's philosophy of life is very like Lucy's. Her beautiful, statuesque daughter mirrors Lucy's daughter Ethel, who had five years earlier also married into a titled family. The dilemmas of the mother and the conflicting advice she is given have the ring of truth about them and it is tempting to make autobiographical conjectures about parts of this novel. All ends happily when a skeleton in his own cupboard is revealed and Sir George becomes less judgmental about the failures of others. Lucy re-used her themes sometimes making big changes. The short story, *A Sad Comedy* appeared in *The Last Touches and Other Stories.* It reappeared as *A Supreme Moment: a play in one Act*, in *Nineteenth Century Review* in 1899. Taking place in the Boulevard Haussman in Paris, the main characters are the same, and the theme – the end of a love-affair – is the same, but the highly dramatic endings are different. In the play the rejected woman draws a gun and threatens, but eventually lacks the courage, to shoot her English diplomat lover. For the stage version the French actress, Madeleine Dubray, tricks the Englishman into declaring his love for one last time as he unknowingly embraces her dead body after she has committed suicide. It is an unusual example of a woman contriving to extract revenge after her death. *Proposals to Kathleen, Being a Maiden's Meditations,* published in 1908 by A.S. Barnes in the USA only, traded somewhat on the 1889 success of *Love Letters of a Worldly Woman.* It was an update of Lucy's 1888 contribution to Temple Bar of *A Chapter on Proposals.* It was well reviewed across the USA, and is a monologue in which Kathleen, on the night before her wedding, reads a bundle of earlier proposals of marriage. Her sharply perceptive comments on the letters are as revealing of the suitors' characters as of her own needs. It was received as: 'A clever little book whose author has a clear understanding of men.' In the end Kathleen accepts, with no great enthusiasm, a safe but dull suitor. 'There will be rest and quiet and safety with him that do not exist anywhere else for me . . . and sometimes I feel that all this longing after the intellectual life and its amenities is but a form of longing for human companionship and sympathy, perhaps love, that I may be proud to know is mine.' she reminisces about former suitors before she burns all the letters and prepares to face up to life with her chosen partner.

In June 1893 Lucy's first one-act play *An Interlude* was produced at Terry's Theatre, and three years later *A Honeymoon Tragedy* appeared at the Comedy Theatre, but her first really successful foray into the theatrical world was in October 1900 with *The Likeness of the Night.* This play attracted a lot of attention and was reviewed at length in newspapers across the country from October 1900 through till the autumn of the next year. There were three reasons for the intense attention. First, it was a most dramatic 'problem' play. Secondly, before production, it had been the subject of a controversy that raged for some time in the London press, and thirdly, it was most beautifully staged and produced by the Kendals who took two main parts in it. Dame Madge Kendal and her husband, at that time one of the most famous actor/producer couples, took up Lucy's play after it had been rejected by Mr George Alexander some years earlier. The extraordinary coincidence was that the first six nights of the Kendals' first production of it at the Royal Court Theatre in Liverpool, ran parallel with the last six nights of George Alexander's production of *A Debt of Honour* – a play with a marked resemblance to *The Likeness of the Night* – at the St James's Theatre in London.[36] Lucy had accepted, after a public exchange of letters in the press, that *A Debt of Honour* was not a plagiarism. She also accepted that Sidney Grundy, the author, had not seen her magazine story *The End of her Journey* that had preceded it in 1885 in *Temple Bar,* or the publication of the full text of the play a few months earlier. Nevertheless the whiff of scandal meant that almost every review of *The Likeness of the Night* mentioned the controversy. The general view was that the only similarity lay in their common theme of a man with two women. Simon Grundy arranged for the wife to kill the mistress, but Lucy's self-effacing, saintly wife, committed suicide in order to leave her husband free. Unmitigated woe following sexual duplicity was Lucy's theme, and Nemesis comes in the last lines when the real reason for the death of the first wife is revealed to the now married mistress. 'Keep Back! Keep Back! – between us flows the sea!', is her melodramatic cry to her new husband, and both of them know that there can be no further happiness for them. Lucy chose lines from Swinburne to illustrate the terrible shattering of dreams when the truth is revealed to the lovers:

> And where the red was, lo, the bloodless white,
> And where the truth was, the likeness of a liar,
> And where day was, the likeness of the night;
> This is the end of every man's desire.[37]

After mixed receptions in Glasgow, Brighton, Liverpool, and Manchester it came to the Grand Theatre, Fulham, where it was a huge success. A year later it was successfully revived, with a certain amount of cutting, at the St James's Theatre. It ran for sixty-three performances and took up the whole of Clement Scott's 'First-Night' review in the *Telegraph.* He wrote:

> It is not often that a piece is greeted with such genuine, unaffected and overwhelming cheers as those which were heard last night . . . coming in sudden irresistible volume from the hearts of deeply moved spectators.

The acting contributed in no small way to the success of the play. Clement Scott wrote: 'Mrs Kendal gave a perfect lesson in the whole art and craft of stage technique . . . she gave a wonderful, clear-cut, and artistically proportioned picture of a wife with a broken heart.' Lucy visited Vienna after this success and arranged for the translation into German by Dr Kellner. The director of the Raimund Theatre accepted it but although it ran as a *feuilleton* in the *Wiener Neues Tagblatt* it was never produced there. The short story from which the play was derived, *The End of her Journey*, was chosen by Kate Flint for her 1996 *Oxford Anthology of Victorian Love Stories.* However Lucy could never repeat this triumph. Her plays *A Woman Alone, Two's Company, Eve's Lover* and *The Latch* were presented on stage with only moderate success. The London production of *Hamilton's Second Marriage* at the Court Theatre in 1907 was reviewed as disappointing and unconvincing.

Lucy's love for France came through also in a play that she wrote in 1901 while she was renting a 'pavillon' in St Germain, near Paris. Her story *The Long Duel* had first appeared in *Strand* Magazine 1891. It was reputed to be based on an incident in the life of Meissonier the famous French painter. She adapted the story as a play, *The Last Touches*. In the short-story, the famous painter Carbouche is tricked by the woman he had loved from his youth, into painting a flattering portrait to satisfy her vanity. Her final cruelty is to laugh at him, take the portrait and leave him with a handful of banknotes. In the play the mistress again tricks the painter into altering his portrait, which shows her as old and hard. But, in the process of flirting with him, she realises that she still loves him and they end in each other's arms. Frederic Harrison praised her for 'wonderful realism and truth and most convincing actuality . . . [catching] the tone of the Paris *ateliers.* The motif is so French, the dialogue so Parisian . . . . Is it impossible to get it put into fine French and tried on a French stage?' Lucy later wrote:

This play when it appeared in *Fortnightly Review* attracted the attention of Sarah Bernhardt. She had it read to her in French by her secretary (P. Rockham) who can testify how much she liked it. Sarah said she would like to do it if Coquelin would play Carbouche. Coquelin said he would gladly read it if it were translated but he did not like hearing a play read. It was translated, but he unkindly died just as it was ready to show to him, and the translation still sits in my drawer.

However, the play definitely did not appeal to Henry James:

Dearest Lucy C, I have waited to welcome you, to thank you for your dear and brilliant Vienna letter . . . . I've believed that not till now (if even now), would you be disengaged from the tangled skein of your adventures. And even at this hour (of loud-ticking midnight stillness) I don't pretend to do more than greet you affectionately on the threshold of your home; promise you a better equivalent (for your so interesting, so envy-squeezing, so vivid record of adventure) at some very near date; and, above all, renew my jubilation at your having made so good and brave a thing of it all . . . . I am just emerging from a domestic cyclone . . . . My two old man-and-wife servants (who had been with me sixteen years) were, a few days ago, shot into space (thank heaven at last!) . . . (and I am living blissfully . . . with a house-maid and a charwoman, and immensely enjoying my simplified state and my relief from what I now see was a long nightmare.) I read your play in the *Nineteenth Century*, as you invited me, but I can't write of it now beyond saying that I was greatly struck by the care and finish that you have given it. If I must tell you categorically, however, I don't think it a scenic subject at all; I think it bears all the marks of a subject selected for a tale and done as a play as an after-thought. I don't see, that is, what the scenic form does, or can do, for it, that the narrative couldn't do better . . . . And here I am douching you on your doorstep with cold water. We must talk, we must colloquise and compare and renew the first moment we can, and I am all the while and ever your affectionate old friend, Henry James.[38]

The play never appeared on the stage except for a one-act adaptation in French, which appeared at the Odéon in Paris without Lucy's knowledge or permission, and with no reference to her as the author. John Lane published an edition in 1902 and Lucy again tried to interest Henry James by sending him a copy in 1909. But his view had not changed:

A play appears to me of necessity to involve a struggle, a question (of whether, and how, will it or won't it happen? and if so, or not so, how and why? – which we have the suspense, the curiosity, the anxiety, the tension, in a word, of seeing; and which means that the whole thing shows an attack upon oppositions – with the victory or failure on one side or the other, and each wavering and shifting from point to point.) But your hero . . . is passive, he doesn't take the field . . . . Don't hate me for saying these things – for working them out critically, and so far as may be, illuminatingly, in the face of the difficulty the L[ong] D[uel] seems to have had in getting itself brought out. We are dealing with an art prodigiously difficult and arduous in every way – and in which one seems most of all to sink into a Sea of colossal Waste. I'm not sure that The Other House, after all my not to-be-reckoned labour and calculation on it, isn't (to be) wasted. But these are dreary words – it is much past midnight. I am damned critical – for it is the only thing to be, and all else is damned humbug. But I don't mean to be a douche of cold water, and am ever so tenderly and faithfully yours, Henry James.[39]

As with Lucy and *The Long Duel*, Henry James had written *The Other House* as a short story many years before, and was reworking it for the stage. It too was never produced.

Harpers of New York published Lucy's story *Margaret Vincent* in 1902, but the book came out in London at the same time as *Woodside Farm*. This is an interesting philosophical novel dealing with a clergyman who loses his faith and is, as a result, jilted by the woman he loves. Escaping to the quietness of the Surrey countryside, he marries a widow and has a daughter. As this child grows up and goes out into the world, his former life catches up with them and the book follows the changes that knowledge of the past bring to the lives of the main characters. The Lucy-like daughter/heroine valiantly tackles conventional religious and social prejudice armed only with simple truth, an open mind, and a love of the world. The book engages the reader because of the way the very different personalities of the powerful women in it are observed and drawn, and, of course, in spite of contrived and unlikely turns of events, the daughter wins through and gets her true love in the end.

Again, a foreign setting added interest to *The Getting Well of Dorothy* published by Methuen in 1904. Lucy wrote this book while staying at Clarens on the lake of Geneva. It chronicles day-to-day incidents in the lives of the two little girls 'Dorothy and Betty', who are clearly Lucy's daughters, Ethel and Margaret. It was reviewed well by *The Standard* as a 'fascinating little book' full of keen observation of childhood experience and illness. It was translated into Dutch and titled *Dora en Betty*. However we get a different view of it from Lucy's literary agent, A.P. Watt, who gave his opinions to Scribner in 1892 when Lucy had first tried to get it published in America. He wrote, 'Mrs C's reputation is brilliant, but *The Getting Well of Dorothy* is not!!' The book is a mother's touching record of her children growing up, and a reminder of the travels that Lucy made with William before he died. A thinly disguised Leslie Stephen visits them, buys a silver chain for 'Dorothy', and is recognisable in one of the twenty-six illustrations by Gordon Browne. Again, there is the autobiographical element. Mr Murray the husband does not appear – he is a soldier and far away, and is an echo of William. 'He had blue eyes and a voice that not only made people do anything he told them, but made them feel how wise and happy they were to do

*One of the few existing photographs of Lucy.
Unfortunately this one is undated.*

it. Everyone loved him, and to have known him even a little was something to be thankful for all one's life long.' The book's appeal is largely limited to mothers, though it ran to four editions, but to those with children it rings true, and Lucy's inimitable observation of simple events makes moving reading. In her diary in the 1920s, Virginia Woolf was to ridicule Lucy for keeping, in her sitting room, 'a wooden black cat on the clock, and little carved animals under it'. From *The Getting Well of Dorothy* we understand why such baubles, given to her by her children, were treasures to Lucy.

Of her later novels, only *Sir George's Objection* and *Miss Fingal* registered any critical interest, but *The House in Marylebone* (1917) is rather intriguing since, although she never conceded it, it may be based on her 'lost' early years. It follows, with autobiographical flavour, the experiences of a Lucy-like girl who leaves home to live alone and work for her living as a typist in London. She travels alone in Europe, settles into a house in Marylebone with a group of independent career girls and establishes for herself many of the opinions and attitudes that are strongly present in Mrs W.K. Clifford herself. The book, published by Duckworth, sold well and appeared in the cheap editions of the Standard Collection of British and American Authors.

As happened with Mrs Humphry Ward, Lucy's appeal had begun to decline in the early twentieth century. She never regained her early readership, and her later years were a rather sad battle to interest publishers and producers. Although she did not take the open stand that Mrs Humphry Ward had done when she became leader of the Women's Anti-Suffrage League, Lucy nevertheless aligned herself with those who did. In her *New York Times* interview during her visit to America in 1912, when she was directly asked if she was a suffragette, she answered vehemently, 'No!' Her stance on the question of suffrage was to 'Sit on the fence, and if either side ask me I say, "Please go away or I will fall down on the other side." 'She was reported to hold the view that people who pay taxes should have the vote regardless of sex, and expressed the reactionary opinion that voting rights should only go to those who had the capacity and character to merit it. Lucy claimed that she had been wildly misquoted in the article but naturally enough, this sort of stance would strike her dead as far as the younger reader was concerned. Lucy, who had struggled valiantly, even deviously, to appear 'young', was not politically agile enough to move with the times and depict her heroines as sympathetic supporters of the 'new woman'. Instead she blew the trumpet on her own demise as she proudly stated that, as far as the suffrage question was concerned, she had, 'never written a single line about it' in her any of her books.

In June of 1905 Lucy's daughter Ethel married Fisher Wentworth Dilke, the eldest nephew of Sir Charles Dilke. Ethel had already established herself as a poetess with regular publication in newspapers and journals, and had two published volumes to her name.[40] Lucy took perhaps more pride in Ethel's successes than in her own. By then Lucy was earning enough money from her own writings to give up her fourteen-year association with *The Standard*. She wrote that she 'had bigger literary fish to fry' than reviewing other people's books, and, with one of her daughters safely married to a relatively wealthy man, she was able to spend more time 'frisking' in Europe. She did however contribute a short story to the 1905 anthology of poems, stories essay, drawings and music, brought out by the *Daily Mail* and dedicated to the Queen.[41] Her story *Concerning Breadsnatchers* was a tribute to her husband's philosophy of life and it took its place with contributions from W.W. Jacobs, Bram Stoker, Marie Corelli, Hall Caine, Edward Elgar, Holman Hunt and Arthur W. Pinero.

The 1919 publication of *Miss Fingal* was Lucy's last real success and in one aspect her most unusual book. In it Lucy sought to portray the mystical idea that the spirit of a dead person could, in certain circumstances, take possession of a living one. Lucy created a bland, passive, spinster heroine in order for her to be transformed when the secondary character dies. The colourless Miss Fingal of the early part of the book becomes involved with Linda, an unhappily married and seriously ill mother of two lovely children. At the exact moment of Linda's death Miss Fingal is involved in an accident which renders her unconscious. In some psychic way, the dying woman's personality enters Miss Fingal's unconscious body. The results of this ostensible reincarnation dominate the second half of the book. It was an ambitious proposition, was widely reviewed and had a mixed reception. Many critics judged it to be her finest novel although some readers missed the full implication of the transformation of Miss Fingal. Lucy later sought, without success, to dramatise it.

Lucy's position as literary critic for *The Standard* must have influenced her choice of '*Critically London*' for the telegraphic address of her Colville Road home. Sir Arthur Conan Doyle, who was a close friend of hers, chose for his Psychic Bookshop's telegraphic address: '*Ectoplasm, Sowest London*'! When in 1919 Lucy sent him a copy of *Miss Fingal* he remarked that her book showed her 'wonderful power of character drawing to be a strong as ever'. He later wrote to her suggesting that she look into the 'psychic question', inviting her to one of his lectures, and offering to lend her books. He felt that he was 'overdrawn on the bank of life' and commented about spiritualism: 'that way assurance and happiness lie . . . all fear of death is lost . . . . Why man should refuse or neglect such enormous consolation is a mystery to me.' [42] But Lucy, remembering, perhaps, her husband's and Moncure Conway's active contempt for séances, was not tempted.

While she was writing *Miss Fingal*, Lucy admitted to losing confidence about the psychic elements she had introduced. She decided to send the proofs to her friend, the great literary scholar and Professor of English at London University, William Paton Ker, and ask his advice. W.P. Ker was the particular friend of the travel writer Freya Stark, and he fell dead in 1923 while climbing in his beloved Alps with her. Lucy also consulted another of her closest friends, Sir Sidney Colvin, who had just retired as Keeper of Prints and Drawings at the British Museum. Both men expressed their strong approval of the book and its unusual theme and encouraged her to go ahead. She mentioned the support she had received from these people when she sent the book to Scribner and was hugely embarrassed when their names were published to advertise the book in America. She sent distraught messages across the Atlantic and had the names removed from later advertisements. Charles Scribner apologised and admitted that a junior editor had bungled. In all of her writings, the freshness comes from the introduction into her stories of realistic details gathered from her travels abroad and authentic elements from the London life that she and William had enjoyed. The lack of any formal religious constituent in her stories reflects her husband's influence upon her and she drew upon the personalities of her eminent friends to create some of her characters. The moving and poignant elements come from her interpretation of the emotional needs of women in their relationships with the men they love. It is true that she was attracted to morbid themes, but she used them to provide deep insights into the hearts of her characters. Her skill in these portrayals produces in the reader flashes of recognition and response, which can transcend the often stilted conventional settings and contrived dénouements that 'date' and diminish the appeal of some of her work today. Dr Marysa Demoor, writing in *Turn of the Century Modernism and Modernity in Literature and the Arts* considers it strange that Mrs W.K. Clifford is not remembered at all today. She writes:

> In my opinion she is clearly another representative of a group of turn-of-the-century women novelists (nowadays generally unknown) who – in the words of Gerd Bjørnhovde – were developing 'non authoritarian' narrative techniques. Sarah Grand, George Egerton,
> Margaret Harkness and Olive Schreiner, according to Bjørnhovde, all seemed to employ the same fictional strategies, which allowed them to back away from a traditionally male authority. These devices – the elusive narrator, the disintegrating plot and the images of transition – were part and parcel of Mrs Clifford's writing technique.

Lucy published just one book, *George Wendern Gave a Party*, under the pseudonym John Inglis. She did this in 1912, simultaneously with Scribner in the USA and *Blackwood's Magazine* in London. It is hard to understand her motive, for she was established and respected as Mrs W.K. Clifford. The record of her decision to do so seems extraordinarily casual. She wrote to Scribner: 'Choose a man's name, John Inglis or James Brooks or anything you please.' Her friend Mrs Humphry Ward wrote:

> I was much interested in 'John Inglis's' book! I remember well reading some of it in Blackwood and wondering who John Inglis was. The financial plot and the chief men characters are extremely vigorous and alive and the gathering of the threads towards the end most ingenious. I particularly admired the scene between the lady housekeeper and her sister at the beginning – excellent! Pray – are you going on as John Inglis or do you relapse into Mrs W.K. Clifford again?[43]

Lucy did not repeat her experiment with pseudonyms in England, though two stories, *The Chance* and *Gavin's Wife,* she sent to *Scribner's Magazine* as by John Inglis. Although Scribner admired *Gavin's Wife* as being a pleasure to read and 'ingenious and quite different from her other stories' they decided they were not able to publish it because they were running a serial by Edith Wharton at the time. Henry James also noted Lucy's pseudonym and joked, after complaining about mediocrity in the theatre: 'What a public, what critics and what a grotesque world – if it wasn't for dear John Inglis! – who becomes my Uncle now!' The last remark of course was because she was his 'Aunt Lucy'. She set the scene for *George Wendern* among big business syndicates and sharp-dealing entrepreneurs, but, her usual theme of the troubled paths of lovers runs strongly through the story.

In many of Lucy's novels there is a thinness of descriptive writing and Henry James provided perhaps the most pertinent criticism of her work in a letter he wrote in May 1912 when Lucy had just returned from her visit to New York. He began with his thoughts about his own sense of antiquity after having had his voice described as 'old' when reading his paper on 'The Novel in *The Ring and the Book*' at the Royal Society of Literature. He was sixty-nine, and, although she claimed to be younger, Lucy was sixty-six:

> Dearest Lucy,
> Your impulse to steep me, and hold me down under water in the Fountain of Youth . . . is no less beautiful than the expression you have given it, by which I am more touched than I can tell you. I take it as one of your constant kindnesses . . . . The young, on the whole, make me pretty sad – the old themselves don't. But the pretension to youth is a thing that makes me saddest and oldest of all; the acceptance that I am all the while growing older decidedly rejuvenates me . . . .So I am young enough – and you are magnificent, simply; I get from you the sense of an inexhaustible vital freshness, and your voice (so beautiful!) of your twentieth year. Your going to America was admirably young – an act of your twenty-fifth. Don't be younger than that; don't seem a year younger than you do seem; for in that case you will have quite withdrawn from my side. Keep with me a little . . . . I find G(eorge) W(endern) brisk and alive. But I have to take it in pieces, or liberal snips, and so have only reached the middle. What I feel critically (and I can feel about anything of the sort but critically) is that you don't squeeze your material hard and tight enough, to press out of its ounces and inches what they will give. That material lies too loose in your hand – or your hand, otherwise expressed, doesn't tighten round it. That is the fault of all fictive writing now, it seems to me – that and the inordinate abuse of dialogue – though this is but one effect of the not squeezing. It's a wrong, a disastrous and unscientific economy altogether. I squeeze as I read you – but that, as I say, is re-writing! However, I will tell you more when I have eaten all the pieces. And I shall love and stick to you always – as your old, very old, oldest old H.J.[44]

In the January 1920 edition of *Bookman,* Wilfred L. Randell, the literary critic, briefly defined the undertone that runs through all of Mrs Clifford's writings as, 'the beauty of the world, the joy of living and the happiness of doing well whatever task life has set before you'. Lucy's photograph adorned the front cover and the three-page review of her books was highlighted along with *The Poetry of Hardy* and *The Poetry of Kipling*. Her death was noted in *Bookman* 1929 and the article contained these words:

> Her work seems lately to have fallen into neglect which is undeserved, though probably inevitable in these overcrowded days: but, when most of the crowd are gone past, it seems probable that two or three of her books will be found remaining with us yet for many years.

It would be appropriate if – now that research and access to her correspondence has cast light on her remarkable life – more of her writings could be readily available.

Notes

1. *Oxford Book of Modern Fairy Tales* 1993; *Oxford Anthology of Victorian Love Stories.* 1997; *Nineteenth Century Short Stories by Women*, edited by Harriet Devine Jump, Routledge 1998;*World Wide Web Wonderbook*, 1998, short story 'The Bad Girl' (from Anyhow Stories).
2. Mrs W.K. Clifford, *The Dingy House in Kensington*, Cassell, Petter, Galpin and Co., 1881.
3. Virginia Woolf, *The Diary of Virginia Woolf*, Introduction by Quentin Bell, editor. A.O Bell, five vols, Hogarth Press, 1977-1984, Vol 2, Entry for Sat. 24 Jan, 1920.
4. Notably Adeline R. Tintner, *Intimacy Between James and Lucy Clifford*, English Literature in Transition, 1880-1920. 44:1 2001, pp. 106-110.
5. ALS (one of two) Valehouse Collection.
6. Mrs W.K. Clifford, *Anyhow Stories: Moral and Otherwise*, A and C Black, 1882.
7. Mrs W.K. Clifford, *Anyhow Stories for Children*, Duckworth 1885.
8. Alison Lurie, *Don't Tell The Grown-ups. Subversive Children's Literature*, Bloomsbury, 1990, pp. 67-73.
9. victoria@iubvm.ucs.indiana.edu. *19th Century British Culture and Society*, 6 October, 1996.
10. F.J. Harvey Darton, *Children's Books in England*, Cambridge University Press, 1932, p.203.
11. Robert Lee Wolff, *Nineteenth Century Fiction, A Bibliographical Catalogue*, Vol. 1.
12. Gillian Avery and Angela Bell, *Heroes and Heroines in Children's Stories*, Hodder and Stoughton, 1965, p.52.
13. Jonathan Cott (ed.), *Beyond the Looking Glass. Extraordinary works of Fairy Tale and Fantasy*, Hart-Davis and MacGibbon, 1974.
14. Rudolf Ekstein, *American Imago: A Psychoanalytic Journal for Culture and Science and the Arts*, 1978, Vol. 35, pp. 124-145.
15. Alison Lurie, *A Tail of Terror,* New York Review of Books, December 1975, p. 26.
16. David G. Hartwell (ed.), *The New York Review of Science Fiction,* has featured Lucy Clifford's work and included *The New Mother* in his anthology of horror stories, *The Dark Descent,* 1987.
17. Heather Schell, Clifford's 'The New Mother' and the Menace of the Lower Classes, *Turn of the Century Women*, Department of English, Georgetown University, Washington, DC. Vol. V, Number 1&2, 1990, pp. 41-47.
18. Lucy Clifford's papers, Valehouse Collection.
19. *Mrs Keith's Crime*, T. Fisher Unwin, 1887
20. ALS (one of four), Valehouse Collection, and one of the ten letters to Mrs Clifford included in *The Collected Letters of Thomas Hardy*, edited by R.L. Purdy and M. Millgate, Clarendon Press 1978, Vol. 6, p. 116
21. *Church Times*, 15 December 1885.
22. ALS, (one of four) Valehouse Collection.
23. ALS, (one of three, and part of a fourth) Valehouse Collection.
24. ALS (one of thirty-nine), Valehouse Collection.
25. ibid.
26. ALS, (one of three), Valehouse Collection.
27. ALS (one of seven), Valehouse Collection.
28. As note 15.
29. Letters from Lucy Clifford held at the British Library, Add. MSS 44, 521 f.134: Two L.C. letters to W.E. Gladstone are in this collection.
30. ALS, Valehouse Collection.

31. Inscribed book held in Valehouse Collection.
32. *The National Review,* July 1893. Later published by W.H. Allen in their 'Three-and-Six penny' series along with seven stories by other writers.
33. Mrs W.K. Clifford, *A Woman Alone*, Methuen, 1901.
34. There are 297 communications from Lucy to the American Publisher Charles Scribner's Sons held in Rare Books and Special Collections, (CO101) Box 32, Princeton University Libraries, USA
35. Mary Angela Dickens, *A Chat with Mrs W.K. Clifford.* Windsor Magazine, 9, 1899, pp. 483-485.
36. A.E.W. Mason, *Sir George Alexander and the St James' Theatre*, Macmillan, 1935, pp. 140-143
37. A.C. Swinburne, *Ballad of Burdens,*
38. M. Demoor and M. Chisholm, *"Bravest of women and finest of friends": Henry James's letters to Lucy Clifford.* English Literary Studies, University of Victoria, 1999, Item 11
39. ibid, item 41.
40. Ethel Dilke, *Songs of Dreams,* 1903. and *Love's Journey,*1905. Both volumes published by John Lane, Bodley Head, and dedicated to her Mother.
41. *The Queen's Christmas Carol*: *An Anthology*, Published by the *Daily Mail*, 1905.
42. ALS (one of three), Valehouse Collection.
43. ALS (one of seven), Valehouse Collection.
44. As note 38, Item 43.

The Clifford Heritage

This Chapter was written jointly with my husband, Professor R. Chisholm

To every reasonable question there is an intelligible answer, which either we or posterity may know by the exercise of scientific thought.[1]

In his lifetime, William Clifford was recognised as a world-class mathematician and was famed for his lectures and publications on science and philosophy. But this fame did not last. While his work was not completely forgotten after his death, it is only in the last quarter century that his name has again become well known amongst mathematicians and scientists.

A principal reason for this revival is that one of Clifford's main mathematical ideas has been seen to have widespread importance in mathematical physics and in several branches of mathematics. In two mathematical papers, William Clifford defined what are now known as 'Clifford Algebras'.[2] Until around 1970, only a handful of mathematicians and theoretical physicists had studied this work. Since that time, Clifford Algebras have become widely known and used, regular conferences have been held, and a number of books have been published on the subject. Two questions come to mind immediately. What are Clifford algebras, and why are they important? Why, for about a hundred years, was the subject generally ignored?

H.J.S. Smith, the editor of Clifford's *Mathematical Papers*, says that William Clifford was 'above all and beyond all a geometer'. This accords with Clifford's geometrical view of numbers. In the section entitled *Steps,* in the first chapter of *Common Sense of the Exact Sciences,* Clifford explains how ordinary numbers can be seen as distances stepped along a 'number line'. Positive numbers correspond to steps in one direction and negative numbers to steps in the opposite direction. Clifford strongly emphasised this picture of numbers as movements on a line. Our 'ordinary' space is three-dimensional, and so it is important to be able to describe geometry and motion in this space. William Clifford approached this problem by asking: 'How can we extend the idea of steps on a number line to motions in space?' At that time, William Rowan Hamilton had already used the term 'vectors' to describe movements in space.[3] Nowadays, we can see 'wind vectors' on weather maps on television, pointing in the direction of the wind; small arrows for light winds, big arrows for strong winds. Anyone with experience of navigation of boats knows that the velocities of a boat in the water, and of the water itself, are drawn as arrows whose length represents speed; and they know that the true velocity is given by completing the 'triangle of velocities' with these arrows as two of its sides. This process is known mathematically as 'vector addition'. Hamilton had invented a mathematical scheme based on vectors, which described all geometric concepts, such as areas and volumes, in three-dimensional space. In space, there are three quite distinct directions which are perpendicular to each other, like those at the corner of a room, and Hamilton had discovered an algebraic rule to describe perpendicularity of two vectors. Incorporation of this rule into his scheme, which he called 'quaternion algebra', completed the description of three-dimensional space.

William Clifford wanted to do more than Hamilton had done. He wanted to describe all possible motions of solid bodies travelling and rotating in space. In 1873, he published a long paper describing an extension of Hamilton's quaternion algebra.[4] This new mathematical scheme turned out to do a great deal more than Clifford had originally intended, and he eventually realised that he had invented 'an algebra' describing four-dimensional space.

It is important to explain what this means. In quaternion algebra, there are at most three vectors which satisfy Hamilton's *algebraic* rule for describing the fact that they are perpendicular to each other. That is why quaternion algebra corresponds to space of *three* dimensions. Clifford's new scheme contained *four* objects which were perpendicular to each other according to Hamilton's *algebraic* rule. In mathematical language, this new scheme is said to *define* 'four-dimensional space' and the four objects are interpreted as vectors. This simply means that the three-dimensional *algebraic* structure has been extended.

Clifford then realised that this was only the first step in an unending chain. He could extend this mathematical scheme in a simple and systematic way to include any number of vectors that were 'perpendicular' to each other in this *algebraic* sense. The mathematical scheme with, say, twenty vectors satisfying Hamilton's condition is now known as the Clifford algebra of twenty-dimensional space. We should not conclude that Clifford, or anyone else, could visualise a space of twenty dimensions. Where our geometric intuitions fail us, the systematics of algebra take over.

It is always hard to explain why some discoveries are not immediately appreciated by the scientific world. One reason for the neglect of Clifford algebra was that the idea of 'higher-dimensional spaces' was not fashionable in the nineteenth century. Indeed, some contemporaries of Clifford argued strongly that we live in a three-dimensional world, and that it was meaningless to talk of spaces of four, five, and higher dimensions.

Another difficulty was that there was a great deal of argument even about Hamilton's quaternion algebra, and two very eminent and influential academics opposed its use. Arthur Cayley, professor of mathematics at Cambridge and one of the finest mathematicians of the nineteenth century, used other methods and saw no need for quaternion algebra. William Thomson, Lord Kelvin, a leading physicist of the age, said that quaternions, 'though beautifully ingenious, have been an unmixed evil to those who have touched them in any way'.[5] He refused even to use the word 'vector'. Although the argument continued, Cayley and Lord Kelvin outlived the main proponents of Hamilton's methods, William Clifford and James Clerk Maxwell, who both died in 1879. Meanwhile, Willard Gibbs in New Haven, Connecticut, had produced notes for his students on a simple 'algebra of vectors', very effective for our three-dimensional space. His is the everyday practical technique now used world-wide by mathematicians, scientists and engineers. Several twentieth-century discoveries showed the relevance of Clifford algebras to various branches of mathematics and physics. Perhaps the most dramatic development was the entry of Clifford algebras into the theory of fundamental particles. In 1928, Paul Dirac published his revolutionary theory of the electron, a theory which rapidly became a corner-stone of all modern theoretical physics. To describe the electron mathematically, Dirac invented a mathematical structure known as 'Dirac algebra'. He did not realise that this algebra was very closely related to Clifford's algebra describing four-dimensional space, discovered half a century earlier. It was another twenty-five years before physicists began to appreciate that the algebra Dirac had invented was just a particular one of a whole range of Clifford algebras.

The coincidence of the close relationship between Clifford's and Dirac's algebras is important. Remember that each Clifford algebra corresponds to its own 'space'. Dirac's algebra is precisely the Clifford algebra associated with the 'space-time' of Einstein's Special Theory of Relativity – the four-dimensional union of ordinary space and time which appears to be fundamental to our physical universe. So, out of a myriad of possible mathematical structures, *the same Clifford algebra turns out to describe both space-time and the fundamental particles of nature.*

William Clifford did not know of space-time and Relativity; nor did he know about the electron, discovered eighteen years after his death. But in his review of *The Unseen Universe*, he did conjecture a close relationship between 'the ether', the mysterious background to the universe, and 'molecules', which were then the shadowy basic constitutions of matter. He wrote:

> Until, therefore, it is absolutely disproved, it must remain the simplest and most probable assumption that they are finally made of the same stuff – that the material molecule is some kind of knot or coagulation of ether.[6]

If we modernise this conjecture, replacing 'material molecule' by 'electron', and 'ether' by 'space-time', we have the suggestion that space-time and the electron 'are made of the same stuff'. What we now know is that they are described by the same Clifford algebra, but we do not yet see them as 'the same stuff'. Had Clifford lived a full life span, perhaps he could have solved this problem for us!

The close relationship of Clifford algebras to fundamental physics was exposed in the Maryland lectures of Marcel Riesz in 1959 and in David Hestenes' book *Space-time Algebra*. Riesz emphasised the geometric properties of Clifford algebras; Hestenes concentrated more on the relationship of the Dirac algebra to elementary particle theory.

Several other quite different and important developments were taking place in the 1960s. Michael Atiyah and his collaborators showed that Clifford algebras were fundamentally related to a modern form of geometry called 'topology'.[7] This is the study of the structure of things which may have a number of holes in

A cartoon image of the 'Clifford corkscrew', by Dave Chisholm, cartoonist for the Sunday Times, *as described by Frederick Pollock in the Introduction to Clifford's* Lectures and Essays.

them, and which can be deformed without being broken – the simplest example is a rubber quoit or a rubber band; each of these have one hole in them, and they are 'topologically equivalent'. Roger Penrose used a Clifford algebra to define 'twistors', which have been used in a variety of mathematical and physical contexts.[8] R. Shaw, H.W. Braden, U. Dempwolff and others have studied the abstract structure of Clifford algebras, and have demonstrated their close relationship to certain 'finite geometries'.

A quite different development had been started in the 1930s by R. Fueter and his collaborators. A very beautiful mathematical theory based on 'complex numbers' – in which we invent and use a 'square root of minus one' – had been developed over a century earlier, and could be interpreted in terms of geometry in two-dimensional space. Fueter's group showed that Hamilton's quaternions could be used to extend this theory to three dimensions. In the 1960s, Richard Delanghe and David Hestenes independently showed that, by using Clifford algebra, this theory could be extended to any number of dimensions. At about this time, two academics with great foresight began to apply Clifford algebras to engineering problems: Georges Deschamps who was fascinated by the symmetry of the Dirac algebra; and Folke Bolinder, a student of Marcel Riesz, who applied Clifford algebra to network theory. All of these areas of research are still being developed, and a number of computer programmes have now been written which allow Clifford algebras to be used in a wide variety of tasks in mathematics, science and engineering. Practical applications include robotic design and tomography, which enables tumours to be located from measurements on the exterior of the human body. The subject has come of age.

Clifford's algebras corresponded to what we call 'flat spaces', but he did not define algebras corresponding to 'curved spaces'. However, the extension of his algebras to curved spaces has been carried out using various mathematical formalisms; in particular, Ruth Farwell and Roy Chisholm have established a direct and uncomplicated way of doing this. Had he lived, Clifford might well have generalised his ideas to curved spaces, since he was extremely interested in the concept of curvature: he translated from German Riemann's doctoral thesis, which gave a general mathematical definition of curved spaces.

'Flat space' and 'curved space' are mathematical concepts which need some explanation. Let us first discuss 'flat space'. The floor of a room is 'flat'; a mathematician would call it a 'flat two-dimensional space'. He would also call the three-dimensional space of our everyday intuitions 'flat' or 'Euclidean', extending the meaning of flatness. In the third of his lectures on *The Philosophy of the Pure Sciences,* given at the Royal Institution in 1873, Clifford spelt out in detail the properties of a flat space.[9] The essence of flatness lies in two assumptions. One is that an object does not change shape if it is moved around in the space. This is something that we take for granted every moment of our lives: things do not change shape when we take them from the kitchen into the dining room. The second essential property is that a figure can be enlarged or diminished without any change of shape. Enlarging a photograph exemplifies this second assumption. If we use old-fashioned methods, light is spread out uniformly by lenses to produce the enlarged image, which is the same shape as the original photograph.

We are so familiar with these very obvious properties of 'ordinary' space that we find it hard to imagine any other situation. So it is very difficult to imagine the effects of curvature of three-dimensional space, in which things would necessarily be distorted if they were moved or enlarged. However, we do have experience of 'curved two-dimensional spaces'. One example is the curved surface of an armchair. If we try to cover an armchair with a length of cloth – which is flat when we lay it on a table – we cannot do it without cutting and gathering the material in order to make it fit – it is not 'the right shape'. If we tried instead using a very elastic material, we might cover the chair simply by stretching the material. Then we would be 'changing the shape' of the material, in order to make it fit the curved shape of the armchair. Also, sliding the elastic material over the curved surface would necessitate it changing shape. The essence of 'curvature' is the need for things to change their shape.

Human intuition does not go far beyond the image of the two-dimensional curved surface of a range of hills, or the covering of an armchair. To a great extent, mathematicians working on curved spaces rely on calculations to produce their results. The geometry of flat space was laid down by Euclid and other Greek mathematicians. Newton's laws of motion, based on Euclid's geometry, were for 150 years the main basis for understanding the physical world. So deep is the intuition of flat space, and so successful was Euclidean geometry, that it became a fundamental philosophical tenet. For centuries, mathematicians and philosophers argued about the possibility of alternatives to flat Euclidean space. Kant, in particular, held that God had implanted certain *a priori* understandings in our minds; among these was the intuition of Euclidean space, so that we were unable to comprehend any other possibility. The argument was resolved in the nineteenth century, to a great extent by Riemann's broad definition of curved spaces.[10]

William Clifford was strongly influenced by Riemann's work, and a detailed account of this connection has been given by Ruth Farwell and Christopher Knee.[11] Clifford made several conjectures about curvature of physical space. The first was the subject of a short paper to the Cambridge Philosophical Society in 1870.[12] He wrote:

> Riemann has shown that there are different kinds of lines and surfaces, so there are different kinds of space of three dimensions; and that we can only find out by experience to which of these kinds the space in which we live belongs . . . . I wish here to indicate a manner in which these speculations may be applied to the investigation of physical phenomena. I hold in fact:
> (1) That small portions of space are in fact of a nature analogous to little hills on a surface which is on the average flat; namely, that the ordinary laws of geometry are not valid in them.
> (2) That this property of being curved or distorted is continually being passed on from one portion of space to another after the manner of a wave.
> (3) That this variation of the curvature of space is what really happens in that phenomenon that we call the motion of matter, whether ponderable or ethereal.
> (4) That in the physical world nothing else takes place but this variation, subject (possibly) to the laws of continuity.

This prophecy once again reflects Clifford's view that 'particles' of matter are just a peculiar part of the structure of space. Up to the present day, there is no convincing fundamental theory that unifies 'particles' and 'space', so that this conjecture remains as a possible clue to a future 'grand unified theory'.

Another of Clifford's conjectures is often quoted as an anticipation of Einstein's General Theory.[13] In it he talks of triangles drawn in space, but he recognises that, in curved space, there are no such things as 'straight lines'. There are, however, 'straightest possible lines' – the simplest example being the lines of longitude drawn on the curved surface of a globe of the world. If we draw several triangles on a flat piece of paper, and measure the three angles in each triangle, we shall find that their sum is always 180 degrees, or two 'right angles'. This is a property of Euclidean space, valid for everyday life. Clifford questioned whether very large triangles might be distorted by curvature of space on an astronomical scale:

> I am supposed to know that the three angles of a rectilinear triangle are exactly two right angles. Now suppose that three points are taken in space, distant from each other as the Sun is from Alpha Centauri, and the shortest distances between these points are drawn so as to form a triangle . . . then I do not know that this sum would differ at all from two right angles; but I also do not know that the difference would be less than ten degrees.

The General Theory of Relativity assumes that large masses cause a curvature of space, and is based upon Riemann's mathematics. Einstein's theory now agrees with a great deal of experimental evidence, and explains, for example, how the gravitational forces of the sun, the earth and the moon can be interpreted as curvature of space-time produced by these large masses. It replaces and slightly modifies Newton's law of gravitation, which is not quite accurate. Of course, our intuitions favour Newton's view, that the earth's force of gravity pulls everything downwards in our flat Euclidean space. A cricketer catching a high ball would not be helped by the comment that, apart from air resistance, the ball was simply following the straightest possible path in curved space-time.

Another crucial test of Einstein's theory concerns rays of light. He made the assumption that they will pass along the 'straightest possible' paths in space. The mass of the sun causes nearby space to be curved. So light from a star travelling towards the earth will be deflected if it passes close to the sun. Einstein calculated this deflection, and his prediction was tested during the eclipse of the sun in 1919. The result supported Einstein's theory. Nowadays astronomers are able to observe some astronomical objects at far greater distances, and to measure the deflection of their light by whole galaxies. A few observations have been made of a single object producing two images, one on each side of a galaxy: the light is bent towards the galaxy as it passes round each side, and so appears to come from two different directions. This is one of the most convincing demonstrations of the curvature of space. Joan Richards, in *Mathematical Visions,* investigates the whole history of geometry in Victorian times. She notes that, in his discussion of space, Clifford 'emphasised Spatial experience rather than perception'. In other words, Clifford explains curvature of space by imagining objects being moved around and thereby being forced to change shape. So 'objects' and 'space' are, for him, inextricably linked – a view reinforced by his assertion that particles are simply 'little hills' of curvature in space. Arguments about the nature of space and matter have gone on down the ages. Lucretius expressed the view of Democritus and the Greek atomists, 'there are bodies and there is void in which these bodies are placed and through which they move about'. Parmenides and Melissus held the opposite view that the universe was one continuous unchanging whole. Melissus argues, 'Nor is anything empty, for the empty is nothing and that which is nothing cannot be'. The remarkable success of Newton's mechanics confused the argument, since it was based on the axioms of absolute space and time, roundly criticised by Leibnitz and others. But Newton also laid down a principle with which Clifford would not have quarrelled:

> Therefore geometry is founded in mechanical practice and is nothing but that part of universal mechanics which accurately proposes and demonstrates the art of measuring.

It was left to Ernst Mach, preceding Einstein, to point out that Newton's 'absolute space' should be replaced by the physical concept 'at rest relative to the background of stars'.

The idea that 'empty space' and 'objects in space' are physically distinct still lingers in modern physics. Even in General Relativity, we think of mass causing, and responding to, curvature of space. Farwell and Knee remind us that Riemann himself thought that geometry was separate from physics. Clifford's view of the inseparability of space and matter is an integral and distinctive part of his 'monistic' philosophy.

To Clifford, the possibility of space being curved was of great philosophical importance. First, it exploded the idea that physical space was inevitably Euclidean flat space, given *a priori* according to Kant; the nature of space on very small and very large scales had to be determined by experiment. Clifford insisted that our knowledge of space and time will always be finite, so that we can never establish scientific laws which are 'at an infinite distance', implying that we can never know that physical space is Euclidean, or which are true 'for ever'. A second limitation of our knowledge, spelt out in his essay *On the Aims and Instrument of Scientific Thought*, is that all measurements are inexact – however carefully we measure, there is always a margin for error.[14] So we can never say that an observed law of nature is 'exact', and we must be prepared to modify even very well-established laws.

These scientific principles brought Clifford into conflict with James Clerk Maxwell, for whom he had great respect, over the establishment of accurate and uniform standards of measurements, which were increasingly needed in science, engineering and commerce throughout the nineteenth century. Simon Schaffer has detailed how metrology had become entangled with religious and political prejudices, and with superstitions about the Great Pyramid.[15] Maxwell dispensed with standards based on the dimensions of the earth or its orbit, or those of the Pyramid, and very reasonably proposed using the wavelength of a line in the sodium spectrum as a basic standard of length. He and William Huggins had shown that this wavelength was the same on earth as in the spectra of stars, but Maxwell himself gave a theological interpretation of this observed uniformity. He followed William Herschel in believing that atoms and molecules were 'manufactured' and were eternal permanent structures, providing exact standards for measurement of mass, length and time. These eternal manufactured articles required, of course, a 'Manufacturer' who had laid down the building blocks of the Universe. Maxwell's views were attacked by both Thomas Huxley and John Tyndall, but the most telling criticism came from Clifford in *The First and Last Catastrophe*.[16] His objections were essentially the same as those to the assertion that space was Euclidean. He writes:

> If there is any name among contemporary natural philosophers to whom is due the reverence of all true students of science, it is that of Professor Clerk Maxwell. But if anyone not possessing his great authority had put forward an argument, founded apparently upon a scientific basis, in which there occurred assumptions about what things can and what things cannot have existed from eternity, and about the exact similarity of two or more things established by experiment, we should say 'Past eternity; absolute exactness; this won't do'; and we should pass on to another book.

Joan Richards remarks that Clifford's mathematical and philosophical ideas have often been treated separately, and makes the important point that he saw mathematics, science and philosophy as parts of a unified 'world view':

> In many of his writings, however, these two aspects of the man [mathematical and philosophical] are intimately connected. The new view of geometrical truth, which was involved in the non-Euclidean ideas of the metrical geometers, was an integral part of Clifford's arguments for scientific naturalism. In his writings he clearly spelled out the implications of the new geometry in the highest context of English post-Darwinian discussion.[17]

The cohesion of his ideas is evident in *The Aims and Instruments of Scientific Thought,* in which he spelt out all the main principles of scientific investigation, insisting that these principles should govern the whole of human thought. He talks about the deduction of the existence of the planet Uranus, of engineers building bridges, of biological classification, of a grocer weighing sugar, of the geometry of space, of Stanley's telegraph message that he had found Dr Livingstone, and many other varied examples. The application of Roman Law, the words of a poet, and the principles of a moralist, are all, for Clifford, an application of scientific thought, which is 'not an accompaniment or condition of human progress, but human progress itself'. In making these claims for the methods of science, Clifford was explaining and developing the philosophy of 'scientific naturalism', following in the steps of Spencer, Huxley and Tyndall: they all shared Clifford's wide perspective of human activity and experience. Lightman and Young note that this breadth of scope was shared by many of the opponents of scientific naturalism, and that the arguments encompassed not only scientific, philosophical and religious questions, but also economic, political and social issues.[18] The background to scientific naturalism during the first half of the nineteenth century, when William Paley's *Natural Theology* became the orthodoxy of the Anglican Church and of Oxbridge, is detailed by Turner and in the essays edited by Helmstadter and Lightman.[19]

This orthodoxy was by no means unchallenged: Turner focuses on the apostasy of Newman's Roman Catholic tendency, Darwin's explanation of evolution by natural selection, and Ruskin's insistence that art should be a personal expression of truth. Alternative religious views were held by Comte, Anglican Evangelicals and various nonconformist groups. Hennell and Strauss questioned the literal truth of the Bible.[20] But the most fundamental challenge to religious orthodoxy came from scientific books such as Lyell's *Geology* and Chambers' *Vestiges of the Natural History of Creation.* The broad advance of scientific knowledge suggested that science might be able to explain away all the mysteries of life and the universe. The increasing evidence that the chemistry and physics of living and inert matter were fundamentally the same all provided challenges to Anglican orthodoxy. The *Bridgewater Treatises*, written in the 1830s by a number of eminent scientists, sought to reconcile natural theology with the advances of science. Their aim was to proclaim through scientific knowledge, 'the Power, Wisdom, and goodness of God, as manifested in his Creation'[21] But these treatises did not answer the fundamental theological problems of pain and evil, and signalled that theological dogma had increasingly to give ground to scientific discovery and understanding. Geological evidence of the antiquity of the world had to be accepted, and the fixity of living species had to be abandoned in favour of adaptation arranged by the Creator.

The final hammer-blow, which eventually forced the Anglican establishment to reconsider its whole position, was the theory of natural selection put forward in 1858 by Wallace and Darwin, and expounded a year later in Darwin's *Origin of Species*. Their idea, crudely described as 'survival of the fittest', provided a natural mechanism for the adaptation and evolution of living organisms. No longer was there a need for a Creator and Adaptor of life, and men were now related to monkeys, not to angels. This new natural philosophy needed a prophet, and that prophet was Thomas Huxley, known as 'Darwin's bulldog'. The controversy over science and religion became very public when he declared, in answer to the taunt of Samuel Wilberforce, his willingness to accept that he was descended from a monkey. Huxley, a master of written and spoken language, publicised Darwin's ideas widely in articles and numerous informal lectures, criticising the obscurantism of the Church. Other eminent proponents of the new naturalism were Joseph Hooker, Director of Kew Gardens, who had helped to lambast Wilberforce at Oxford, John Tyndall, who succeeded Faraday as Director of the Royal Institution in 1866, and Charles Lyell, who published *The Antiquity of Man* in 1863. Huxley also led the way in insisting that our beliefs must be based on scientific method – he referred to science as 'organised common sense' – and attacked 'the host of weaker men whose sense of truth has been destroyed in the effort to harmonise impossibilities – whose life has been wasted in the attempt to force the generous wine of Science into the old bottles of Judaism'.[22] The resignation of Leslie Stephen from his Tutorship at Trinity Hall in 1862, on the grounds that he could not continue to preach a religion in which he had never really believed, brought the controversy to a head in Cambridge, and it was in the ferment of this discussion that William Clifford arrived with his High Church beliefs. Turning against these beliefs cost him a great deal of heart-searching; but he was an outspoken religious atheist by the time he was elected to a Trinity College Fellowship in 1868. This was the same year in which a majority of the Fellows of Trinity petitioned parliament to liberalise the Statutes, and by 1871 the Gladstone government was compelled to remove all religious requirements from conditions of appointments at Cambridge.

At the time, this triumph of secularism and anti-clericalism must have seemed to many to be a catastrophic break with the past – deposing God was pretty drastic. But the argument did not end there, and indeed has not yet ended. Also, Lightman and Young see natural religion as a compromise; one stage in the progression from a belief in the literal truth of Genesis, to the acceptance of the scepticism of scientific naturalism.[23] Huxley, Tyndall and Clifford all sought some form of natural religion to replace formal religions, and it may well be that, in his less polemical moments, William Clifford would have agreed that scientific naturalism had evolved through natural theology. He saw our individual lives as part of the stream of human life flowing down the ages, and our thoughts and discoveries as a contribution to the continuing process of development of human learning and morality. He was generous in acknowledging the influence of many scientists and philosophers on his own thinking – but he always reserved the right to disagree!

For Clifford, all true knowledge was scientific, and so had to be, like scientific laws, an understanding derived from our experience. A scientific law is a particularly precise form of understanding, based on the results of experiments carried out under clearly specified conditions. The law is then a prediction that a certain pattern of results will *invariably* follow, given these same conditions. Clifford insisted that no law could ever be proved absolutely; but that, if the experiments are repeated many times in different places, and the results are *always* consistent with the prediction, the law may be accepted, at least provisionally. Any discrepancy in an experiment must be explained, perhaps by a failure of the apparatus. But if there is no sensible explanation, the experimenter must accept that he has not in fact found a true 'law of nature'. What he must *not* assume is that 'nature' has been capricious. In several of his lectures and essays, Clifford emphasised that all of science is based on the assumption of a 'uniformity of nature'.[24] In *The Ethics of Belief*, he discusses the example of the 'bright lines' of light in the spectrum of hydrogen in laboratory experiments. When these same characteristic spectral lines are observed in light from the sun, we assume that they show that hydrogen is present in the sun, and has the same properties as hydrogen on earth. This was precisely the empirical uniformity used by Clerk Maxwell to establish a universal basis of measurement; but as we have seen, while Clifford agreed with this as a sensible practical procedure, he profoundly disagreed with Maxwell's dogma that the uniformity of nature arose from the Creation of identical eternal atoms. For Clifford, uniformity of nature was a common sense hypothesis which enabled us to formulate scientific laws, and, more generally, to coordinate our views of the world at different places and different times. He pointed out that we rely on uniformity of nature in everything that we do, even when the observed uniformity is specialist expertise accumulated by others. We trust a doctor to prescribe and treat us because his experience and understanding of the nature of disease allow him to assume the likely outcome of the treatment he recommends. This may seem to be stating the obvious, but even today many people are ready to believe that supernatural powers can be called upon to overrule natural laws or to effect miracles of healing.

Astrological belief, that heavenly bodies can dictate personality and bring good or bad luck, is even more prevalent. We still need people like Clifford and Conway to expose these irrational and fraudulent beliefs, and to insist that these 'magical' happenings are either delusions or clever trickery. To Clifford, miracles were impossible, since they would be isolated breaches of the uniformity of nature and of the dictates of common sense. If there were a God who could arbitrarily suspend the laws of the universe he had created, how could we hope to understand the universe rationally?

Clifford argued that, since science and 'everyday common sense' are based on the same principles, scientific thinking should be applied to every 'practical question'. This included questions of religion and ethics. The teaching that scientific principles of thought should be applied universally was central to his mould-breaking essay *The Ethics of Belief*, in which he made his uncompromising affirmation that 'It is wrong always, everywhere and for anyone, to believe anything upon insufficient evidence.'

Since Clifford taught that science and 'common sense' went hand in hand, it is reasonable to ask why he should have argued that space might be curved. After all, Euclidean geometry, together with Newtonian mechanics, gave a very accurate picture of the world, and was the basis for understanding modern technology and the planetary system. Clifford did not argue that Euclidean geometry was wrong for practical purposes, for ordinary scales of distance, but he did not like the idea of space 'going all the way out to infinity' in the way that Euclid and Newton had ordained. To him it was philosophically more satisfactory for space to be curved on a large scale, so that it might turn in on itself and be finite. Also, the assumption that space was curved on very small scales enabled him to envisage a way of unifying space with particles of matter. Clifford realised that this departure from the 'common sense' view of nature was counter-intuitive, and emphasised that on very large or very small scales of space and time, nature might be quite different from the world of our immediate experience. The theories of Relativity and the Quantum theory have shown that Clifford's caution was fully justified. Earlier we mentioned Clifford's disagreement with Kant: this went deeper than an argument about whether space is curved or not. They both acknowledged the fact that, at a very early age, humans develop a remarkable intuition and understanding of the three-dimensional space we live in – for example, the co-ordination of sight, hearing and touch which develops in the first few months of life. They totally disagreed about the cause of this intuition. Kant, remember, thought that God had implanted the *a priori* intuition of Euclidean space in our minds. Clifford, who was well versed in physiology, took the Darwinian line: over perhaps a thousand million years, natural selection had led to the development of more and more complex creatures, with increasingly sophisticated nervous systems.[25] In complex life forms, including humans, these nervous systems have become particularly adapted to developing intuitions about the space we live in.

This was only one of many ways in which the evolutionary theory challenged old theological beliefs. The observation of the natural development and improvement of complex life forms, particularly mankind, over many ages, had made nonsense of the dogma of the creation of separate fixed species by 'God the Creator'. Some theologians, for instance Charles Kingsley and James Martineau, sought to reconcile religion with Darwin's theory of natural selection. God was the moving spirit of the Universe, laying down the ground rules of existence and of morality. For Clifford, this was an unjustified assumption, since there was no evidence for the existence of this God. If he were asked 'How did the world begin?' he would reply, 'I do not know. Perhaps someone will find out in a few hundred years.' Admission of ignorance was for him of paramount importance. In *Aims and Instruments of Scientific Thought*, from which the chapter heading quotation is taken, he spells out the meaning of a 'reasonable question', expressing confidence that we, or our successors, can find answers to sensible questions. But in the most poetic of his essays, *Cosmic Emotion*,[26] he emphasises our ever-changing view of the cosmos, not only in space and time, but in all scientific fields, including the development of human nature and morality. Certainly, 'practical questions are within the domain of science', but 'our conception of the universe is for us, and not for our children, any more than it was for our fathers'. So while, in our own time, we can frame and sometimes answer 'reasonable questions', the world moves on in each generation to a new perspective and to new questions. For Clifford, 'Solving the Universe' was a wonderful adventure, but he believed that we shall never come to a final complete understanding of nature. For Darwin, Huxley, Spencer and Clifford, evolutionary ideas accounted for more than the purely physical development of living species. In several of his lectures and essays, Clifford details how moral principles develop as working rules within communities, and become part of our individual instincts.[27] In part we are taught, and in part we inherit, those feelings of 'fairness' and 'social obligation' which we call 'conscience'. That we have a sense of 'right and wrong' does not mean, in Clifford's view, that we are 'born knowing the Sermon on the Mount', but that we are naturally receptive to ideas of justice and social coherence. He accepts that we have an 'inner voice of conscience', but he insists that it is *not* the voice of God – it is the ancestral voice of the 'Tribal Self'.[28] He uses biblical language when he writes:

> The voice of conscience is the voice of the Father Man within us; the accumulated instinct of the race is poured into each one of us.

Like many philosophers over the ages, Clifford thought a great deal about the relationship between 'mind' and 'brain'.[29] The physiological research of Helmholtz and others had begun to explain how the nervous system transmits electrical impulses to and from our brains, and it was known that happenings in our conscious *minds* – sensations, memories, ideas and intentions – were invariably accompanied by electrical disturbances within the physical *brain*. Clifford did not avoid drawing the conclusion that individual life after death was not possible, though he had sympathy with those who, in grief or through fear, took refuge in belief in an afterlife.

Clifford accepted, and sought to explain, this duality between the brain and the individual conscious mind. In his search for a unified view of all experience, he came to the conclusion that the fundamental material of existence was what he called 'Mind-stuff'.[30] However, he warned his audience that they might not understand this idea, and admitted that perhaps he himself did not understand it! Clifford was an idealist in the sense that he accepted that everything we know is based solely on our own sensations and experience, but he still took the common-sense view – assuming, in Huxley's words, 'a good working hypothesis' – that other people's minds exist, and that our idea of an 'external world' depends upon others having consistently similar experiences to our own. The problem is to reconcile the fact that our conscious sensations and feelings are somehow closely dependent upon activity in our brains, which is made of the same atoms as other 'external things'. Clifford took the view that mind and matter were inseparable. Later, Bertrand Russell came to the same conclusion.[31] Farwell and Knee remark that when Clifford was discussing the universe as mind-stuff, 'he was defining space in such a way that representations of the universe and the universe itself are not, in fact distinguishable'. It would certainly be hard to understand how the spatial position of an object in the 'external world' can be reconciled with the position of its image within several different human brains, if this is what Clifford meant. Clifford's warning that mind-stuff might be difficult to understand was certainly justified. Over the ages, philosophers have tried to understand the nature of life. By the mid-Victorian period, chemists had made in the laboratory some of the organic molecules which had earlier been thought to be exclusively 'products of vitality', strongly suggesting that living and dead matter were made out of the same atoms. It was becoming reasonable to suppose that living cells were composed of a particular arrangement of certain molecules, and that a living being was a suitable arrangement of cells. This 'mechanistic' view of life offended many religious people, who felt that it demeaned the handiwork of God. It was also argued that if men were simply complicated machines, we then had no 'Free Will' in governing our own lives.

Argument about the nature of life raged even among scientists and doctors who believed in the chemical basis of living matter. Followers of Darwin strongly held the view that a living creature could only arise from a similar living creature. Louis Pasteur took the same view.[32] In the 1860s, he investigated putrefaction of meat and the growth of other germ colonies. His experiments indicated that food could be contaminated by air-borne germs, but his results were contested by doctors and scientists who believed that germs, which were complex living organisms, could be 'spontaneously generated' from dead tissue in some unexplained way. John Tyndall supported Pasteur, and carried on acrimonious arguments with several medical men. His main opponent was

Dr Bastian, a colleague of William Clifford at University College in the 1870s. Tyndall succeeded, with considerable difficulty, in refuting Bastian's claims that he could 'grow' germs in the laboratory out of sterilised infusions of turnips, old hay and simple chemicals.[33]

Because they believed that living creatures had evolved from simple atoms and molecules, aided by no 'vital force' or spontaneous generation, the supporters of Darwin and Pasteur were accused of 'materialism'. In his highly controversial Presidential Address to the British Association in Belfast in 1874, Tyndall met the criticism head on, accepting the word 'materialism', but spelling out what it meant to him. Yes, living things were composed of the same atoms as the earth and the air. But life was different: it was the product of a thousand million years of evolution – it did not come cheaply. Tyndall entertained the distant possibility of artificial creations of the simplest life forms at some time in the distant future, but made it clear that this was a great gulf yet to be crossed.[34]

Even greater was the gulf between laboratory-produced molecules and complex forms of life, mankind in particular. Tyndall ridiculed the idea that life could arise as a simple chance encounter of atoms. Certainly, in ordinary matter lay 'the promise of all terrestrial life'. But the true story was the gradual evolution of life over the ages, leading eventually to humanity and its 'feelings of Awe, Reverence, Wonder and the love of the beautiful, physical and moral, in Nature, Poetry and Art'. Tyndall, perhaps unwisely, linked modern science with the ideas of Lucretius, in particular with his atomic view of matter, which were then a bone of contention between various religious and philosophical thinkers. This allowed Tyndall's opponents to attack modern atomic theory as old-fashioned, and also to label both Lucretius and the scientific naturalists as irreligious believers in a mechanistic materialism. The convolutions of this argument have been spelt out by Turner.[35] Clifford pointed out the crucial difference between the ancient and modern theories of atomism: Lucretius' theory was a guess, but modern science based its ideas on systematic and thorough experiment. But, like Tyndall, Clifford was prepared to meet the criticism of a mechanistic view of mankind. In *Body and Mind*, he says that an automaton 'goes by itself', and a good automaton responds logically and sensibly to external impulses. He hopes that this does in fact describe his own behaviour. But this is only the external view of a person, and he sees no contradiction with the idea of Free Will, which is our own conscious feeling that, having received various impressions through our senses, we then 'make up our mind' how to respond. Again, in *The Analysis of Mind*, Bertrand Russell takes the same view as Clifford.

William Clifford shared with Tyndall and Huxley their reverence and wonderment at the universe and life. Science, far from destroying wonder at the world by its discoveries, reveals even more wondrous things at every stage. In *Cosmic Emotion*, Clifford writes of those feelings generated by contemplation of the whole vastness of being, of which we comprehend only a small part. This emotion is poetic, and he quotes from Wordsworth, the Odyssey, the Golden Verses from Pythagorean scriptures of the third century, and from two of his favourite poets, Swinburne and Walt Whitman. He also quotes Kant, this time with approval:

> Two things I contemplate with ceaseless awe
> The stars of heaven, and Man's sense of Law.

This matched Clifford's profound concern with understanding the nature of the physical universe and with the innate conscience of mankind. He expressed his high ideals for humanity thus:

> Conscience and reason form the inner core in the human mind, having an origin and nature distinct from the merely animal passions and perceptions; they constitute the soul or spirit of man, the universal part in every one of us.

Clifford was concerned to see the whole of existence as a unity. Man, the most advanced terrestrial life-form, together with his thoughts and feelings, was a part of, and a product of, the physical universe. In his discussion of 'mind-stuff', Clifford expresses a view of matter consonant with that of Tyndall, endowing the simplest particles of matter with a very rudimentary and ill-defined 'quasi-mental fact'. He argues that somewhere in the chain of development of higher life-forms, consciousness gradually began to emerge. So, he deduces, simpler life-forms such as the amoeba must have some vague awareness out of which consciousness develops. But he pushes the argument back further: these simple creatures are made of simpler molecules and these molecules are made of atoms – there is no break in the continuity of being. So the most fundamental particles of matter must be endowed with a very simple basic ingredient of consciousness.[36] The assumption that the elements of consciousness lie within the fundamental particles accentuates the problem of mind-stuff, since, as we have seen, Clifford also thought that particles were some kind of convolution or warping of space itself. Does this mean that space itself is also endowed with some basic quality of consciousness? Had Clifford lived into the age of Relativity and Quantum theory, he might perhaps have related his ideas about mind, matter and space to the very strange properties of elementary particles which we have now been forced to recognise.

Clifford's broad vision of the unity of nature led him to proclaim in *Cosmic Emotion* the existence of life elsewhere in the universe:

> although we have restricted our cosmos to earth in space, and to the history of life upon it in time, there is no necessity to maintain the restriction. For we must suppose that organic action will always take place under the requisite conditions of temperature, light and environment. It is therefore in the last degree improbable that it is confined to our planet.[37]

Only very recently has evidence been found of an abundance of planetary systems attached to other stars. This news would have excited William Clifford.

If we were to ask Clifford which of his ideas he would have liked to understand more fully, he would probably have included the relationship between mind and brain. There would have been an unending number of mathematical problems which he would have liked to solve, and he would no doubt have applied his algebras to a variety of problems in mathematics and physics. Had he lived longer, he would almost certainly, as his dear friend Fred Pollock pointed out, have contributed to the development of the theories of Relativity.[38] He was fascinated by the evidence of a structure of molecules, which he describes as 'at least as complicated as a grand piano' – a very apt simile.[39] He would also have been very excited at the discovery of the electron, and might well have anticipated the role that Clifford algebra now plays in fundamental particle theory.

In William Clifford's private papers is an unpublished note discussing 'revolutions in science', and listing a wide variety of physical and chemical phenomena. The last sentence of this note reads:

> All these things must come out of a knowledge of the form of atoms and their relation to the ether. What is pointed to is therefore a connection between kinetic and undulatory theory.

Was this an anticipation of quantum theory, which unifies the concepts of particles and waves? Sadly, we shall never know what thoughts lay behind Clifford's prophetic conjectures and what understanding of the nature of things he might have given us.

Notes

1. William Kingdon Clifford, *Lectures and Essays,* Vol. 1, Macmillan, 1879, p. 156.
2. LMS Abstract 1876 and later *Mathematical Papers,* 1882, *On the Classification of Geometrical Algebras,* p. 397, and *Applications of Grassmann's Extensive Algebra,* American Journal of Mathematics, Vol. 1, p. 266.
3. W.R. Hamilton, *Mathematical Papers,* Halberstam and Ingram (eds.), Cambridge University Press, 1967.
4. *Mathematical Papers,* 1882, *Preliminary Sketch of Biquaternions,* p.181.
5. Crosbie Smith and M. Norton Wise, *Energy and Empire,* Cambridge University Press, 1989, p. 188.
6. *Lectures and Essays,* Vol.1, *The Unseen Universe,* pp.237-8.
7. M.F. Atiyah, *Collected Works,* Clarendon Press, Oxford, 1985-88.
8. R. Penrose and W. Rindler, *Spinors and Space-time,* Vol. 2, Cambridge University Press, 1986.
9. W.K. Clifford, *Lectures and Essays,* Vol. 1, 1879, p. 295
10. G.F.B. Riemann, '*Uber die Hypothesen welche der Geometrieze Grundleigen'springer,* 1921.
11. R. Farwell and C. Knee, *Stud. Hist. Phil.Sci.*, 21.1.1990.
12. W.K. Clifford, *Mathematical Papers,* 'On the Space Theory of Matter', 1882, p. 21.
13. W.K. Clifford, *Lectures and Essays,* Vol. 1, p. 137.
14. ibid, Vol.1, pp. 124-157.
15. S. Schaffer, *Metrology, Metrication and Victorian Values,* pp. 438-474 of *Victorian Science in Context,* ed. B. Lightman, University of Chicago Press, 1997.
16. W.K. Clifford, *Lectures and Essays,* Vol.1, pp. 191-227

17. J. Richards, *Mathematical Visions*, Academic Press, 1988, p. 109.
18. B. Lightman, *Victorian Science in Context,* Introduction pp. 1-12. R.M. Young, *Darwin's Metaphor*, Cambridge University Press, 1985
19. F.M. Turner, *Contesting Cultural Authority*, Cambridge University Press, 1993. R.J. Helmstadter and B. Lightman,*Victorian Faith in Crisis*, Macmillan, 1990.
20. D.F. Strauss, *'New Life of Jesus'*, trans. G. Eliot, Williams and Norgate, 1865. Among the predecessors whose work Strauss acknowledges is Charles Hennell, a close friend of George Eliot. See C. Hennell, *Inquiry into the Origin of Christianity*, London, 1838.
21. J.M. Robson, *The Fiat and Finger of God: the Bridgewater Treatises,* pp. 39-125 of *Victorian Faith in Crisis*.
22. T.H. Huxley, *Darwiniana*, Macmillan, 1892, pp. 51-52.
23. B. Lightman, *The Origins of Agnosticism*, Johns Hopkins University Press, 1987.
24. W.K. Clifford, *Lectures and Essays*, Vol. 2, *Right and Wrong*, p. 124; *On the Scientific Basis of Morals*, pp. 120-121; *Ethics of Belief,* pp. 205-211; *Cosmic Emotion*, pp. 261-263.
25. ibid, *Body and Mind*, pp. 53-66.
26. ibid, Cosmic Emotion, *Lectures and Essays*, Vol. 2, pp. 253-285.
27. ibid, *On The Scientific Basis of Morals*, pp. 106-123; *The Ethics of Religion*, pp. 212-243, and other essays.
28. ibid, *On the Scientific Basis of Morals*, pp. 109-114.
29. ibid, *Body and Mind*, pp. 31-70.
30. ibid, *On The Nature of Things-in-Themselves*, pp. 71-82.
31. B. Russell, *The Analysis of Mind*, Unwin Brothers, 1921.
32. P. Debré, *Louis Pasteur*, Johns Hopkins University, 1998, Chap. 7.
33. A.S. Eve and C.H. Creasey, *The Life and Work of John Tyndall,* Macmillan 1945, pp. 195-200.
34. ibid, pp. 179-194.
35. F.M. Turner, *Contesting Cultural Authority*, Cambridge University Press, 1993, pp. 262-283.
36. W.K. Clifford. *Lectures and Essays, Vol. 2; Mind and Brain*, pp. 60-61; *On the Nature of Things-in-Themselves,* pp. 82-84.
37. ibid, *Cosmic Emotion*, p. 276.
38. F. Pollock, *For my Grandson*, John Murray, 1933, p. 32.
39. Oliver Lodge writes, 'And as my brilliant teacher, W.K. Clifford, used to say: an atom must be at least as complex as a grand piano.' *Atoms and Rays*, Benn, 1931, 4th edition.

Afterword
The Mathematics of William Kingdon Clifford:

A Personal Relection

The name of William Kingdon Clifford has been familiar to me since my days as a mathematics research student in the early 1950s. He was a geometer of extraordinary talents, with great skills also in algebra and other areas of mathematics. He was, in addition, a philosophical thinker of depth and a wonderful exponent of his ideas in all these areas. There are several independent ways in which his work, achieved in his short life of not quite 34 years, has profoundly influenced the direction of my own research in addition to that of a great many other mathematicians and physicists.

It appears to be accepted that the main mathematical contribution of William Clifford was the introduction of what is now known as "Clifford algebra". This is certainly an algebraic structure of enormous importance, which has also profoundly influenced my own work, as I shall indicate shortly. Yet, this was not Clifford's particular contribution that first made its impression on me when I was a student, nor the one that most manifestly influenced the direction of my later research, its importance growing on me only gradually. To gain an impression of the breadth of Clifford's mathematical work, it may be helpful for me to indicate something of his other contributions that I had earlier been acquainted with.

I think that I first heard Clifford's name in relation to a family of geometrical theorems known as "chain theorems". One of Clifford's circle chain theorems

Figure 1 *A circle chain theorem due to William Kingdon Clifford. NB. the letters "c" (for circles) and "P" (for points) are omitted from the figures (only suffixes given), so as to avoid too much cluttering.*

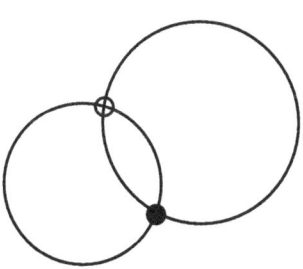 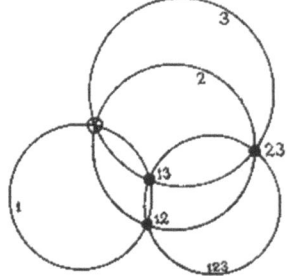

(a) $n=2$: Circles c_1, c_2, through O (ringed point) meet again in P_{12}.

(b) $n=3$: The three points P_{12}, P_{13}, P_{23} lie on the (unique) circle c_{123}.

(c) n=4: The four circles c_{123}, c_{124}, c_{134}, c_{234} all meet at P_{1234}.

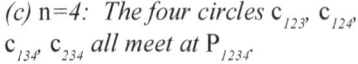
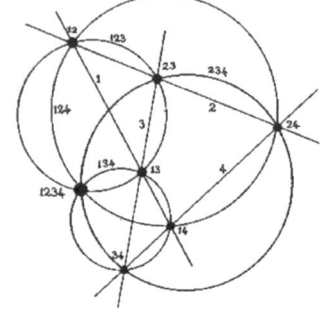

(d) n=4 again, but with O taken to infinity, so that c_{123}, c_{124}, c_{134}, c_{234} are straight lines. This simplified picture applies also to Figs. 1e and 1f, (the original form in which Clifford presented his chain theorem)

can be quite easily explained, and I believe that it is not inappropriate that I try do this here. It concerns the geometry of circles in the ordinary Euclidean plane (although we can even more appropriately think of circles drawn on a sphere). Do not be confused by the word "chain" here. It does not refer to a linking of the circles but to the fact that there is not just a single geometrical configuration giving a single "theorem" in the normal way that one finds in Euclidean geometry, but a "chain of theorems" of increasing complexity, where each one depends upon the one that preceded it. To assert the validity of the chain theorem is to assert that *every* one of this infinite sequence of theorems is true.

To begin, consider two circles c_1, c_2 in general position (on the Euclidean plane or sphere) which share a particular point O, as in figure 1(a). Clearly, they intersect also in a second point, which we label P_{12}. Now consider *three* general circles c_1, c_2, c_3 through O, as in figure 1(b). They will intersect pairwise in three such points P_{12}, P_{23}, P_{13}. There is clearly a circle c_{123} through these three points

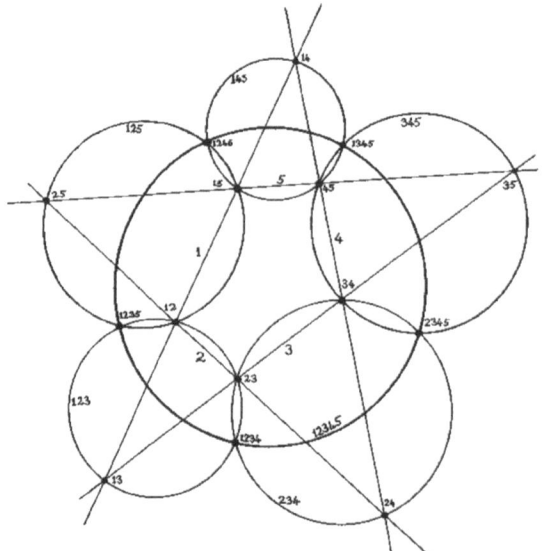

(e) n=5 Miquel's theorem: Start with five general lines c_1, c_2, c_3, c_4, c_5 in the plane. The five points P_{1234} all lie on one circle c_{134567}

Afterword by Sir Roger Penrose 179

(since there is a circle through *any* three points – including a straight line as a limiting case of a circle). Next, as in figure 1(c) and figure 1(d), take *four* general circles c_1, c_2, c_3, c_4 through O. Taking these three at a time we arrive at four final circles c_{123}, c_{124}, c_{134}, c_{234}, by the previous construction. It turns out that these four circles meet in point P_{1234}. This is the first surprise. (This result is actually a direct consequence of an old theorem, known to the ancient Greek geometer (*e*) n=5 *Miquel's theorem: Start with five general lines* c_1, c_2, c_3, c_4, c_5 *in the plane. The 1234·^{p}2345·^{p}1345·^{p}1245 all lie on one circle c12345·* Appollonios, in the 2nd or 3rd century BC.) Now, consider *five* general circles through O, as in figure 1(e). Taking these circles four at a time, we get, by the preceding result, five different final points P_{1234}, P_{1235}, P_{1245}, P_{1345}, P_{2345}. The *new* surprise is that these all lie on one and the same circle (basically a theorem ^{c}12345 due to Miquel). We can now take *six* circles; taking them five at a time, we then find six final circles from he previous construction and, magically, they all pass through one point P_{123456}. With seven circles, as in figure 1(f), we find seven such points lying on one cicle. This continues indefinitely no matter how many circles we start with through O. This is one of the chain theorems that Clifford proved. Of deeper mathematical significance is a remarkable property, discovered by Clifford, of a three-dimensional sphere (that is, a spherical three-dimensional "surface" in four dimensional space). This is the concept of what are now called *Clifford parallels*. These "parallels" are circles that are parallel in the sense that they

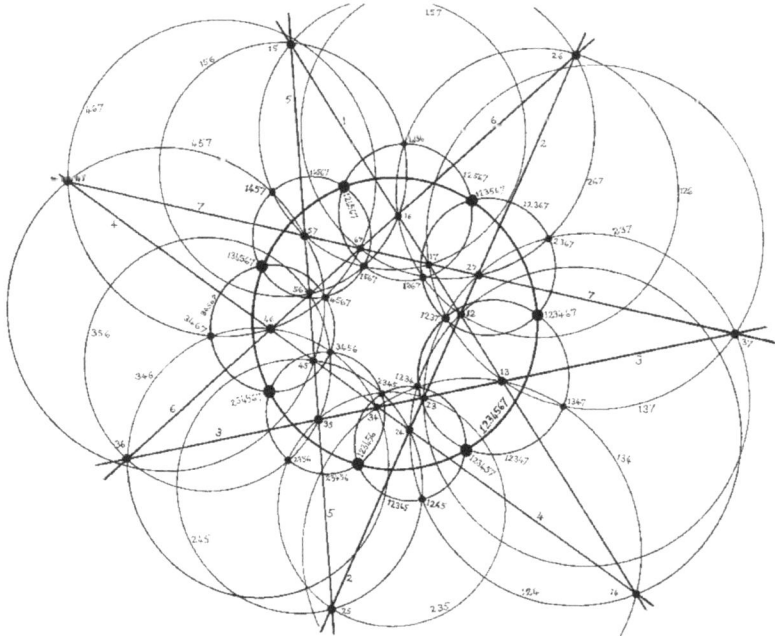

(*f*) n=7 *(representative case): Start with seven general lines* c_1, c_2, c_3, c_4, c_5, c_6, c_7 *in the plane. The seven points* P_{123456}, P_{123457}, P_{123467}, P_{123567}, P_{124567}, P_{134567}, P_{234567} *all lie on one circle* $c_{1234567}$.

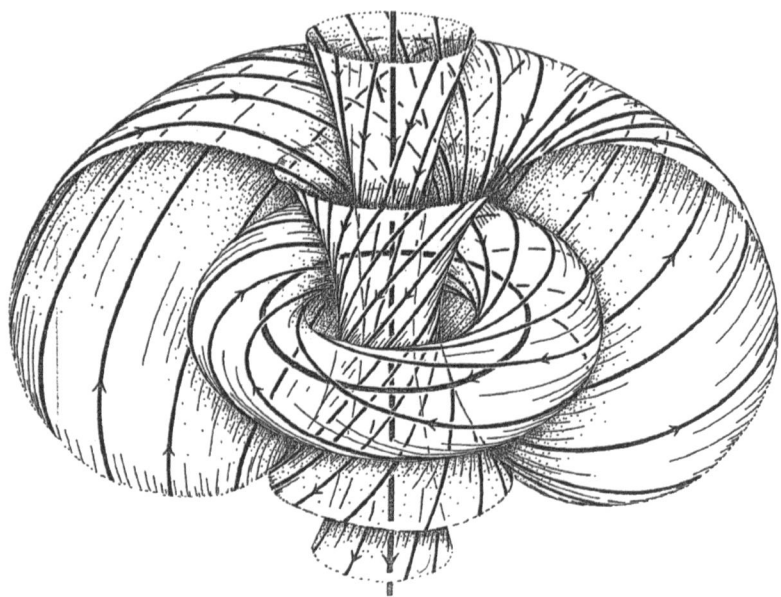

Figure 2. Projection of Clifford's configuration, showing the extraordinary arrangement of twisting and interlocking circles.

never get any closer to each other, nor do they get farther apart, as we proceed along the circles. Yet, the circles are *linked* to one another. Clifford found that the entire three-dimensional sphere can be filled up with such non-intersecting "parallel" circles, each being linked to each of the others. We would now say that the three-sphere is *fibred* by the Clifford circles, and Clifford's construction provides a paradigm of what is now known as a *fibre bundle*.

Mathematical terminology is frequently unfair on the original discoverers of mathematical results, and here we find no exception, for the Clifford fibration of the three-sphere is usually referred to as the now well-known "Hopf fibration"! (In fact, Heinz Hopf was himself scrupulous in giving credit to Clifford, but somehow this credit has got rather lost.)

The geometrical importance of the so-called fibre bundles is something that has emerged in the latter half of the 20th century, and it underlies much of curved-space geometry as well as of the modern theories of particle interactions. Clifford's early example illustrates the essential subtlety of the idea, many years ahead of its time. In addition, Clifford's configuration, when projected into ordinary 3-dimensional space, as in figure 2, provides an extraordinary arrangement of twisting and interlocking circles. This configuration, which turns out to represent the angular-momentum structure of a spinning massless particle, such as the photon (the quantum particle of light), had a big influence on me personally. It provided a geometrical realization of something that I had been searching for, for some years. It launched the subject of *twistor theory* that has been my deepest interest for over 37 years. Our informal publication *Twistor Newsletter* has, for a number of years,

Afterword by Sir Roger Penrose

featured Clifford's configuration on its cover.

It turns out that twistor theory can be approached from various other angles, and one of these is Clifford algebra. Clifford introduced this fundamentally important type of algebra towards the end of his short life, between 1873 and 1878. In 1843, William Rowan Hamilton had discovered the algebra of *quaternions* which provided an algebraic approach to the study of ordinary 3-dimensional space. The novel feature of quaternions is that the so-called *commutative* property of multiplication $ab=ba$ no longer holds, although the other properties of ordinary algebra are retained. What Clifford was able to do was show how Hamilton's algebra could be extended, by combining it with some earlier ideas of Hermann Grassmann, so that it applies to *any* number of dimensions. This is at the expense of the law of *division* (whereby any non-zero element a would have had to have an *inverse* a^{-1}). Clifford's algebra can be thought of as being built up from the basic operation of a *reflection*, an idea that proved to be of great importance in later fundamental work of Élie Cartan, Hermann Weyl and many others in the 20th century. A key role for Clifford algebra appeared in the equation for the electron, discovered by the great physicist Paul Dirac in 1928. In fact, Dirac was unaware of Clifford's much earlier work and, in effect, had to rediscover the necessary part of Clifford's theory for himself. Dirac's equation was accepted as a remarkable breakthrough which represented a turning point in the mathematics of particle physics. It shows an extraordinary foresight on the part of Clifford to have anticipated one of the essential ingredients of this revolution, over 50 years ahead of its time.

Dirac's theory provided a mathematical explanation of the curious quantum-mechanical *spin* of the electron. Indeed, the electron's wave equation is described in terms of what are now called *spinors* (a term coined by the physicist Ehrenfest). Spinors may be regarded as the quantities upon which the elements of the Clifford algebra *act*. Clifford apparently regarded the elements of his algebra simply as entities in their own right, but one can also regard such non-commuting elements as *operators* that act on some other type of entity, in this case spinors. In Dirac's theory, the spinors were *needed*, in order to describe the quantum state of the electron itself, but such a motivation would not have been available to Clifford in his time. The ful ln-dimensional theory of spinors was developed by Richard Brauer and Hermann Weyl in 1935, basing their analysis fundamentally on Clifford algebra. Spinors were first studied, from a different (and more limited) point of view, by Élie Cartan in 1913.

Twistors may be regarded as spinors for *six* dimensions; yet they refer directly to the *four* dimensions of space–time. In effect, when the dimension of the space is increased by two (as here) the spinors double up, so twistors may be represented as *pairs* of space–time spinors. Indeed, one thing that is interesting about the way in which higher-dimensional spinors and Clifford algebras are built up is that the theory for $n+2$ dimensions is, in a sense, a *doubled up* version of the theory for n dimensions. This is curiously analogous to Clifford's chain theorems, where each construction depends upon the one before, the number of circles doubling up at each stage!

Clifford algebras and spinors have acquired an additional importance through the work of Atiyah and Singer, from around 1968. They showed that, in a certain sense, *all* (elliptic) differential equations can be "deformed" and broken down into basic pieces that are essentially the "Dirac equation" in an n-dimensional form. Clifford algebras, and their associated spinors, thereby form a fundamental part of their very general theory. There is a school of thought according to which Clifford algebras should form the basis of a universal approach to physical theory. To my own thinking this viewpoint is somewhat too limited, but I would accept that spinors, in their various guises, do seem to provide us with a deeper perspective on the physical world than the currently more conventional vector/tensor procedures. These can be applied to the curved space–times of general relativity as well as to flat. It was another of Clifford's extraordinarily insightful contributions to realize the importance of Riemann's curved-space geometries to the physical geometry of the universe, quite possibly on the small scale as well as the large. This was some 45 years before Einstein produced his general theory of relativity!

To end on a more personal note, I came across Clifford's posthumous *Mathematical Fragments* about ten years ago, and I was astonished to find that he had been using a diagrammatic notation in the theory of invariants that was very similar to one that I thought I had invented, albeit some three-quarters of a century later! He was extraordinarily inventive in many ways and deeply insightful to the end. But there is no way that one can assess the contributions that this extraordinary man would have made, had he lived to a reasonable age. One has to make do with what he actually achieved. His heritage is, indeed, quite stunning, and we are all his beneficiaries.

Roger Penrose

Bibliography

The bibliography is divided into three parts: the first is of the works of and about W.K. Clifford; the second the works of Lucy Clifford (arranged chronologically); the third a general bibliography.

Works by and on W.K. Clifford

Notebooks, mss and correspondence by and relating to Clifford are mainly held in a private family collection. Original ms for Mathematical Fragments and some letters from period 1871-4 held at University College, London. For details of published works see BL and National Union Catalogue.

Bibliographies

Bibliographical introduction in vol I, *Lectures and Essays* by the late William Kingdon Clifford, FRS, ed L. Stephen and F. Pollock, 1879.

Bibliographical preface in *Mathematical Papers* of W.K. Clifford, ed R. Tucker, 1882.

Collections and selections

Lectures and Essays (hereafter *L&E*) by the late William Kingdon Clifford, FRS, 2 vols, ed L. Stephen and F. Pollock. Vol 1 with an introduction by F. Pollock (biographical, selections from letters, etc and bibliographical), with engraving by C.H. Jeens from a photograph. Vol 2 with a photograph. 1879, London and New York 1886 (1 vol with two essays removed which are published in *Mathematical Papers*), New York 1901 (2 vols), 1918 (Rationalist Press Assoc cheap reprint 52).

Mathematical Fragments (facs of unfinished papers on the theory of graphs), 1881.

Mathematical Papers (hereafter *MP*), ed and preface by R. Tucker (biographical and bibliographical), with lithograph of a page of manuscript written by W.K. Clifford and introduction by H.J. S. Smith. 51 mathematical research papers and appendix containing notes on mathematical topics, lecture notes, syllabuses of lectures, reviews, answers to problems, 1882, New York 1968.

The Scientific Basis of Morals, and Other Essays, New York, 1884.

Conditions of Mental Development, and Other Essays, New York 1885.

The Unseen Universe and Philosophy of the Pure Sciences, New York 1886.

Select Works of William Kingdon Clifford (including *Seeing and Thinking* and other essays from *L&E*). New York 1889.

Der Sinn der Exakten Wissenschaft in Gemeinverstandlicher form Dargestellt von William Kingdon Clifford. Mit 100 Figuren. Deutsche ubersetzung nach der 4. Auflage des englisches originals von der Dr Hans Klemperer. (Translation into German of *MP*), Leipzig, 1913.

The Ethics of Belief and Other Essays, ed L. Stephen and F. Pollock, 1947.

Three lectures on psychology. Reprint of three essays from *L&E* (History of British psychology series, *The Emergence of Psychology*, with an introduction by R. Thomson), 1993.

Books and Tracts

Elements of Dynamic: an introduction to the study of motion and rest in solid and fluid bodies, Part 1 kinematic: Book 1 translations, Book 2 rotations, Book 3 strains, 1878.

Seeing and Thinking: transcript of a series of lectures given in the Town Hall, Shoreditch. (Diagrams by M. Foster). 1879, rptd 1880, New York 1881 (Humboldt Library of Popular Science Literature), London and New York 1890, Cincinatti Teachers' Cooperative Publishing Co, 1891.

Elements of Dynamic: an introduction to the study of motion and rest in solid and fluid bodies, Part 1 (as above). Part 2: Book 4 masses, appendix. (ed posthumously, with a preface, by R. Tucker), 1886.

The Common Sense of the Exact Sciences, Entrusted to R.C. Rowe then ed, prefaced and completed by Karl Pearson, London and New York 1885, second edn London 1886, New York 1888, New York 1894, London 1898, New York 1899, New York 1903, fifth edn London 1907, New York 1946 (new edn with preface by B. Russell and introduction by J.R. Newman, pbd by Knopf), 1947 (Knopf edition pbd by Agent Sigma Books Ltd, London), New York 1955 (reprint of Knopf edn).

Body and Mind, with Other Essays, New York 1891 (Humboldt library of science 145).

Von der Natur der Singe an Sich. Ans dem englischen ubersetztund hrsg. von Hans Kleinpeter, Mit einer ein leitung das heransgebersuber Clifford's leben und werken. (Translation into German of *On the Nature of Things-in-Themselves* with an account of Clifford's life and work), Leipzig 1903.

The Ethics of Religion, New York 1917 (Truth Seeker tracts 130).

The Ethics of Belief and Other Essays, Madigan Timothy J., Prometheus Books, Amherst, New York. 1999.

Contributions to Periodicals

'On Some of the Conditions of Mental Development'. Lecture to and ptd in Proc Royal Institution, 6 Mar 1868, rptd Lectures and Essays Vol 1 1879, rptd *L&E* second edn 1886. (With subsequent letter pbd in *Pall Mall Gazette,* 24 June 1868).

'On Theories of the Physical Forces'. Lecture to and ptd in Proc Royal Institution, 18 Feb 1870, rptd *L&E* Vol 1 1879, rptd *L&E* second edn 1886.

'On the Aims and Instruments of Scientific Thought'. Lecture to the British Association at Brighton, 10 Aug 1872. Reported and rptd in *Macmillan's Magazine* Oct 1872, rptd *L&E* Vol 1 1879, rptd *L&E* second edn 1886.

'Atoms'. Sunday Lecture Soc, 7 Jan 1872, repeated Manchester, 20 Nov 1872, ptd in *Manchester Science Lectures for the people* series 4 no 4 1872, rptd *L&E* Vol 1 1879, rptd *L&E* second edn 1886.

'The Unreasonable'. *Nature* 7, 13 Feb 1873. (Debate with C.M. Ingleby, *Nature* 7, 6 Feb, 13 Feb, 20 Feb 1873).

Rev of Volume 1 of G.H. Lewes *Problems of Life and Mind. Academy,* 7 Feb 1874.

'Atoms'. In D. Estes ed *Half hour recreations in popular science*, Boston 1874.

'The First and the Last Catastrophe', Sunday Lecture Soc, 12 Apr 1874. Ptd (rev) in *Fortnightly Review* Apr 1875, rptd *L&E* Vol 1 1879, rptd *L&E* second edn 1886.

'The Philosophy of the Pure Sciences. Part 1: The Statement of the Question'. Afternoon lectures at the Royal Institution 1 Mar 1873, pbd in *Contemporary Rev* 24, Oct 1874.

'Body and Mind'. Lecture to and ptd in Sunday Lecture Soc 1 Nov 1874, *Fortnightly Rev* Dec 1874, rptd Lectures and Essays Vol 2 1879, rptd *L&E* second edn 1886.

'The Philosophy of the Pure Sciences. Part 2: The Postulates of the Science of Space'. Afternoon lectures at the Royal Institution 8 Mar 1873, pbd in *Contemporary Rev* 25, Feb 1875.

'The Unseen Universe or Physical Speculations on a Future State'. 1875,*Fortnightly Rev* June 1875, rptd *L&E* Vol 1 1879, rptd *L&E* second edn 1886.

'On the Scientific Basis of Morals'. Paper read to Metaphysical Soc 1875, pbd in *Contemporary Review* Sep 1875, rptd *L&E* Vol 2 1879, rptd *L&E* second edn 1886.

'Right and Wrong: the Scientific Ground of their Distinction'. Lecture to and ptd in *Sunday Lecture Soc* 7 Nov 1875, pbd in *Fortnightly Rev* Dec 1875, rptd *L&E* Vol 2 1879, rptd *L&E* second edn 1886.

'Instruments Used in Measurement: Instruments Illustrating Kinematics, Statics and Dynamics'. South Kensington handbook to loan collection of scientific apparatus 1876, rptd *L&E* Vol 2 1879, rptd *MP* 1882.

'The Ethics of Belief '. Paper read to Metaphysical Soc, pbd (with considerable addns) in *Contemporary Review* Jan 1877, rptd *L&E* Vol 2 1879, rptd *L&E* second edn 1886.

'On the Types of Compound S tatement Involving Four Classes'. *Memoirs of the Literary and Philosophical Soc of Manchester*, Session 1876-77, Vol xvi, rptd *L&E* Vol 2 1879, rptd *L&E* 1882.

'The Influence upon Morality of a Decline in Religious Belief '. Pbd in A modern symposium in *Nineteenth Century*, Apr 1877, rptd *L&E* Vol 2 1879, rptd *L&E* second edn 1886, extracts in *Nineteenth century opinion*, comp and ed M. Goodwin, Harmondsworth 1951.

'The Ethics of Religion'. Lecture to and ptd by Sunday Lecture Soc (with title The bearings of Morals on Religion) 4 Mar 1877, pbd in *Fortnightly Rev* Jul 1877, rptd *L&E* Vol 2 1879, rptd *L&E* second edn 1886.

'Cosmic Emotion'. Pbd in *Nineteenth Century*, Oct 1877, rptd *L&E* Vol 2 1879, rptd *L&E* second edn 1886.

'On the Nature of Things-in-Themselves'. Paper read to Metaphysical Soc 1874, pbd in *Mind* Jan 1878, rptd *L&E* Vol 2 1879, rptd *L&E* second edn 1886.

'Virchow on the Teaching of Science', *Nineteenth Century* Apr 1878, rptd *L&E* Vol 2 1879, rptd *L&E* second edn 1886.

'Childhood and Ignorance', *Nineteenth Century,* May 1878.

'The Philosophy of the Pure Sciences. Part 3: The Universal Statements of Arithmetic'. Afternoon lectures at the Royal Institution 15 Mar 1873, pbd in *Nineteenth Century* 5, Mar 1879

'The Philosophy of the Pure Sciences'. Afternoon lectures at the Royal Institution 1, 8, 15 Mar 1873, with the first lecture supplemented by a fragment 'Knowledge and Feeling', pbd in *L&E* Vol 1 1879, rptd *L&E* second edn 1886.

Contributions to collaborative works

The Little People. Stories by Lady Pollock, W.K. Clifford and W.H. Pollock, illus. by J. Collier. Two stories by W.K. Clifford: The New Crown, The Giant's Shoes. 1874.

A Modern Symposium. Ed J. Martineau. Essay by W.K. Clifford: 'The Influence upon Morality of a Decline in Religious Belief'. 1877.

Cosmic Emotion. (With 'The teaching of science' by R. Virchow). New York, 1888.

Letters

L&E. Ed by L. Stephen and F. Pollock (as above), contains a selection from letters written by W.K. Clifford. 1879.

Smith, D.E. 'Clifford's Genius Shown as a Boy'. *Amer. Mathematical Monthly* 29 1922.

Translations

Riemann, G.F.B. 'On the Hypotheses which Lie at the Bases of Geometry' being 'Ueber die Hypothesen, Welche der Geometrie zu Gründe Liegen'. (Habilitationsvortrag, 1854). *Nature* 3 1873.

Reviews and Articles on Clifford and his Works

Anon. 'Professor Clifford's Elements of Dynamic'. *Saturday Rev* 45, 22 June 1878.

Pollock, F. Report on Friday evening discourse, 'Force and Energy by W.K. Clifford' at Royal Institution 28 Mar 1873. *Nature* 22 1880.

Chrystal, G. Review of *Mathematical Papers* and *Mathematical Fragments. Nature* 26, 6 Jul 1882.

Ingleby, C.M. 'Professor Clifford on Curved Space', *Nature,* 7, p.282, 1873.

Madigan, Timothy J., 'Verily, he died too young: A Comparison of W.K. Clifford and Friedrich Nietzsche, *NZ Rationalist and Humanist*, Spring 1998.

Newman, J.R. 'William Kingdon Clifford'.*Scientific American* 188 Feb 1953.

Tait, P.G. 'Clifford's Dynamic (rev of part 1)'. *Nature* 18, 23 May 1878.

Tait, P.G. 'Clifford's Exact Sciences'. Nature 32, 1 1 Jun 1885 (with responses in defence of Clifford by R. Tucker *Nature* 32, 18 Jun 1885, "R" *Nature* 32, 25 Jun 1885, K. Pearson *Nature* 32, 2 Jul 1885).

Pearson, K. 'Elements of Dynamic: Part 1, Book 4 and Appendix'. *Athenaeum* 16 Jul 1887.

Pollock, F (ed with L. Stephen). Introduction in *L&E* by the late William Kingdon Clifford, FRS, 1879 and other edns above.

Power, E.A. 'Exeter's Mathematician – W.K. Clifford, FRS, 1845-1879'. *Advancement of Science* 26 1970.

Works by Lucy Clifford

1871 *The Troubles of Chatty and Molly*, Lucy Lane, serial publication in *The Quiver*.
1872 *About Nellie*, Lucy Lane, serial publication in *The Quiver*.
1873 *Queen Madge*, Lucy Lane, serial publication in *The Quiver*.
1874 *Against Herself*, Lucy Lane, serial publication in *The Quiver*.
1875 *The Bridge Between*, Lucy Lane, serial publication in *The Quiver*.
1876 *Across The Plain*, serial publication in *The Quiver*.
1881 *Children Busy, Children Glad, Children Naughty, Children Sad*, Wells Gardner, Darton and Co., London; T. Nelson and Sons, New York 1881; translated into German.
'Lost', short story published anonymously in *Macmillan's Magazine*.
The Dingy House In Kensington, published anonymously, Cassell, Petter, Galpin and Co. London; G. Munrow, New York. (First published serially 1872 in *The Quiver*).
1882 *Anyhow Stories: Moral and Otherwise*, Macmillan, London.
1884 'From the Top of a Surrey Hill', article published in *St James's Budget*.
'Alassio', article published in *St James's Budget*.
1885 *Mrs Keith's Crime*, R. Bentley and Son, London.
Under Mother's Wing, Wells Gardner and Co.
1886 *Very Short Stories and Verses for Children*, Walter Scott, London.
1887 'Thomas', short story published in *Blackwood's Magazine*.
'The End of her Story', short story published in *Temple Bar*.
'In an Old Chateau', short story published in *Temple Bar*.
'The Last Scene of the Play', short story, published in *Voluntaries for an East London Hospital*, David Stott, London.

1888 'A Chapter on Proposals', short story published in *Temple Bar*.
Anyhow Stories for Children (enlarged edition), Duckworth and Co.
1891 *Love Letters of a Worldly Woman*, Edward Arnold, London. Translated into French and German. (First published serially in 1890 as *Letters of a Worldly Woman* in *Temple Bar*.)
'An Interlude', published in *English Illustrated Magazine*.
'Wooden Tony', published in *English Illustrated Magazine*.
'On the Wane: A Sentimental Correspondence', published in *English Illustrated Magazine*.
1892 *Aunt Anne*, 2 volumes, Harper Bros. New York.
The Last Touches and Other Stories, Macmillan, London.
'A Grey Romance' published in *National Review*.
1893 *A Wild Proxy*, Hutchinson, London.
Marie May or Changed Aims published by Warne. (First published serially in 1887 as *Their Summer Day* in *The Quiver*)
1894 *A Flash of Summer*, D. Appleton and Co. New York, and also in *Illustrated London News*.
1895 *A Flash of Summer. The Story of a Simple Woman's Life*, Methuen, London.
'Luck for Him', short story published in *Cassell's Family Magazine*.
'In Case of Discovery', short story published in *Fortnightly Review*, September edition.
1896 *Mere Stories*, A. and C. Black, London.
A Honeymoon Tragedy, play, publisher unknown.
1897 *The Dominant Note and Other Stories*, Dodd, Mead and Co., New York.
1898 'A Modern Correspondence', published in *Fortnightly Review*.
A Woman Alone, Macmillan, London and New York.
1899 *A Supreme Moment*, play, published in *Fortnightly Review. Likeness of the Night*, published in *Anglo Saxon Review*.
1900 *The Likeness of the Night: A modern play in four acts*, Adam and Charles Black, Macmillan. (Produced Royal Court Theatre, Manchester).
1901 *The Likeness of the Night.*(Produced St James' Theatre London.)
A Long Duel, John Lane, New York and London.
The Searchlight published in *The Nineteenth Century*.
A Woman Alone: Three Stories, Methuen, London.
1902 *Woodside Farm*, Duckworth and Co., London, and as *Margaret Vincent*, Harper Bros. New York.
'The Austrian Tyrol', article published in *The Quiver*.
'At Innsbruck', article published in *The Quiver*.
1903 *The Searchlight*, play, published in *The Nineteenth Century*.
1904 *The Getting Well of Dorothy*, Methuen, London.
'The Way Out', short story published in *The Daily Mail*, London.
1905 'Concerning Breadsnatchers', short story included in *The Queen's Christmas Carol*, An Anthology, *The Daily Mail*.
1906 *The Modern Way, eight examples*, Chapman and Hall, London.
The Shepherd's Purse: A Play, Macmillan and Bowes, Cambridge.
1907 *Hamilton's Second Marriage* (produced Court Theatre, London)
1908 *Proposals to Kathleen, Being a Maiden's Meditations*, Barnes, New York.
The Latch (produced Kingsway Theatre, London)
1909 *Three Plays*, Duckworth, London.
1910 *Sir George's Objection*, Nelson and Sons, London.
'The Chance', short story, published in the August edition of *English Review*.

1912 *George Wendern Gave a Party*, under pseudonym 'John Inglis', published simultaneously by Blackwood, London and Scribner, N.Y. *Two's Company*; *play in Three Acts*, Duckworth, London.
1913 'A Remembrance of George Eliot', article published in the July edition of *The Nineteenth Century*.
1914 *Two's Company* (staged in Manchester). *A Woman Alone*, in three acts, C. Scribner's Sons (produced Little Theatre, London).
1916 *The Searchlight* (produced by Miss Horniman, Manchester)
1917 *The House in Marylebone, A Chronicle*, Duckworth, London.
1918 *Mr Webster, and Others*, Collins, London.
1919 *Miss Fingal*, Blackwood, Edinburgh.
'The Antidote', short story published in *Scribner's Magazine*.
1924 *Eve's Lover and Other Stories*, C. Scribner's Sons, New York.
'Victoria, Lady Welby'. Article contributed to *The Hibbert Journal*, vol. xxiii.
1927 'George Eliot. Some Personal Recollections', article published in the October edition of *The Bookman*.
Eve's Lover, produced at Daly's Theatre, London.
1928 'The Sidney Colvins: Some Personal Recollections', article published in the April edition of *The Bookman*.

General Bibliography

Annan, Noel: *Leslie Stephen: The Godless Victorian*, Weidenfeld and Nicolson, 1948.
Askwith, Betty: *Lady Dilke: A Biography*, Chatto and Windus, 1969.
Bagnold, Arthur Henry: *Shooters Hill Parish Magazine*, 1936-38, Woolwich Libraries.
Beichler, James: *Smoking Guns and Paper Trails*, Unpublished paper on W.K. Clifford. University of Maryland.
Bell, Anne Oliver: (ed.) *The Diary of Virginia Woolf*. Vols 1-5, Hogarth Press, 1977-1985.
Belloc Lowndes, Marie: *The Merry Wives of Westminster*, Macmillan, 1946.
Birkenhead, Lord: *Rudyard Kipling*, Weidenfeld and Nicolson, 1978.
Blind, Matilde: *George Eliot*, W. H. Allen and Co., 1883.
Bosanquet, Theodora: *Henry James at Work*, Hogarth Essays. L. and V. Woolf, 1924.
Bosanquet, Theodora: *Personal Diary 1914-1918*, Unpub. Harvard University Library.
Brandon, Ruth: *The New Women and the Old Men*, Secker and Warburg, 1990.
Clifford, Mrs W.K.: *A Remembrance of George Eliot*, Nineteenth Century Review, 1913.
Clifford, Mrs W.K.: *George Eliot: Some Personal Recollections*, Bookman, Oct.1927.
Clodd, Edward: *Memories,* Chapman and Hall, London, 1916.
Conway, Moncure Daniel: *Autobiography, Memories and Experiences*, Cassell, 1904.
Cosslett, Tess: *The 'Scientific Movement' and Victorian Literature*. Harvester, 1982.
Demoor, Marysa: 'Not with a Bang but a Whimper': Lucy Clifford's Correspondence, 1919-1929, *Cambridge Quarterly*, Vol 30, No 3, 2001
Demoor, M. and Chisholm, M.: *'Bravest of women and finest of friends': Henry James's Letters to Lucy Clifford*, English Literary Studies, University of Victoria, 1999.
Demoor, Marysa: The Women of the Athenaeum in *The Turn of the Century*, Walter de Gruyter, 1995.
Edel, Leon: (ed.) *Letters of Henry James*, Macmillan Ltd., 1974.
Edel, Leon: *Life of Henry James,* Three volumes. Rupert Hart-Davis, 1953.
Farrow, John and Susan: *Madeira,* Hale, 1987.
Gilles, Donald: *Revolutions in Mathematics,* Clarendon Press, 1992.
Gordon, Lyndall: *A Writer's Life*, OUP, 1984.
Greenslet, Ferris: *James Russell Lowell*, Houghton Mifflin, 1905.

Bibliography

Haldane, Elizabeth S.: *George Eliot*, Hodder and Stoughton, 1927.
Hale, E. Everett: *James Russell Lowell and his Friends*, Houghton Mifflin, 1899.
Hart-Davis, Rupert: *Hugh Walpole: A Biography*, Macmillan, 1952.
Howe, Mark de Wolfe: (ed.) *Holmes-Pollock Letters*, Harvard, 1961.
Hyde, H. Montgomery: *Henry James at Home*, Methuen and Co., 1969.
Kipling, Rudyard: *Something of Myself,* Macmillan, 1937.
Krishnamurti, Dr Gutala: *The Eighteen-Nineties*, National Book League, 1973.
Lee, Vernon (Violet Paget): *Letters*, (ed. I.C. Willis) privately printed 1937.
Lubbock, Percy: (ed.) *The Letters of Henry James*, Macmillan, 1920.
Lucas, E. V.: *The Colvins and their Friends,* Methuen and Co., 1928.
Lurie, Alison: *Don't Tell the Grown-Ups: Subversive Children's Literature,* Bloomsbury, 1990.
Macfarlane, A.: *Ten British Mathematicians*, Chapman and Hall, 1916.
Mackenzie, Faith Compton: *William Cory,* Constable London, 1950.
 Extracts, Letters and Journals of William Cory, Pub. by 23 subscribers. Oxford, 1897.
Maitland, F.W.: *Life and Letters of Leslie Stephen*, Duckworth and Co., 1906.
Mallock, W.H.: *The New Republic: Culture, Faith and Philosophy in an English Country House*, Chatto and Windus, 1877.
McGlinchee, Claire: *James Russell Lowell,* Twayne Publishers Inc. New York, 1976.
Newman, James R.: *William Kingdon Clifford,* Scientific American, 188,78. 1953.
Nicolson, Nigel: (ed.) *The Letters of Virginia Woolf,* six vols, Hogarth Press, 1975-1980.
Norton, C. E.: *Letters of James Russell Lowell,* Harper and Brothers, NY, 1894.
Novick, Sheldon M.: *Honorable Justice: The Life of O. W. Holmes Jr.*, Little Brown, 1989.
Nowell-Smith, Simon: *The Legend of the Master*, OUP, 1947.
Pinney, Thomas: *Collected Letters of Rudyard Kipling,* Macmillan, 1990.
Pollock, Sir John: *Time's Chariot*, John Murray, 1950.
Richards, Joan: *Mathematical Visions: The Pursuit of Geometry in Victorian England*, Academic Press, 1988.
Scudder, Horace E.: *James Russell Lowell. A Biography*, Houghton Mifflin, 1901.
Seymour, Miranda: *Henry James and his Literary Circle,* Houghton Mifflin, 1989.
Seymour-Smith, Martin: *Rudyard Kipling,* Queen Anne Press, 1989.
Sladen, Douglas: *Twenty Years of my Life*, Constable and Co., 1914.
Spalding, Frances D.: *Vanessa Bell,* Harvest /HBJ USA, 1983.
Stephen, L. & Pollock, F.: (eds.) *Lectures and Essays:* W.K. Clifford, Macmillan, 1879.
Tucker, Robert: (ed.) *Mathematical Papers:* W.K. Clifford, Macmillan, 1882.
Walpole, Hugh: *The Apple Trees: Four Reminiscences*, Golden Cockerel Press, 1932.
Wolff, Robert Lee: *Sensational Victorian. The Life and Fiction of Mary Elizabeth Braddon*, Garland Publishing Inc., 1979.

Index

Academy, The, 34, 44
Adam Bede, 1
Advancement of Science, 28
Alan's Wife, 82
Alexander, George, 148
Allen, C. Grant, 41
Allen, Jessie, 88, 98
American Imago, 138
Apostles, The, 20, 21, 34
Apple Trees, The, 7
Aquinas, Thomas, 18, 67
Arnold, Matthew, 54
Aspern Papers, The, 145
Athenaeum, The, ix, 9, 39, 71, 120, 129, 146
Athenaeum, The [Club], 77, 83
Atiyah, Sir Michael, xi, 161
Atlantic Monthly, The, 96, 105
Bagehot, Walter, 35
Balestier, Caroline, 76
Barbados, 8
Barrie, J.M., 130
Bastian, H.C, 173
Beerbohm, Sir Max, 91
Belgravia, 53
Bell, Clive, 94, 116, 119
Bell, Florence, 82
Bell, Vanessa, 94, 111, 113, 116, 118, 119, 120, 121
Belloc Lowndes, Marie, 73, 76, 82
Bennett, Arnold, 3, 89, 126, 127, 145
Benson, E.W., 138
Bentley, George, 11
Bentley, Richard, 139, 142, 143, 144 145
Bernhardt, Sarah, 150
Besant, Sir Walter, 142
Beyond the Looking Glass. Extraordinary Works of Fairy Tale and Fantasy, 138
Biglow Papers, 105
Biographical History of Philosophy, 43

Birrell, Augustine, 7, 128, 130
Blackwood's Magazine, 155
Blandy, Charles R., 66
Blandy, Frances Anna, 66
Blind, Karl, 123, 124
Blind, Matilde, 44, 70
Bodleian Library, x, 13
Bodley, George Frederick, 13
Bodley, Sir Thomas, 13
Bolinder, Folke, 163
Boltzmann, L., 2
Bookman, 44, 117, 156
Bosanquet, Theodora, 2, 80, 92, 98, 100
Braddon, Mary Elizabeth (Maxwell), 125, 134
Braden, H.W., 162
Bravest of Women and Finest of Friends Henry James's Letters to Lucy Clifford, 3
Bridgewater Treatises, 150
British Association, 16, 30, 33, 34, 39, 66
Brooke, Rupert, 24, 91
Broughton, Rhoda, 88, 98, 123, 134, 142
Browne, Sir James Crichton, 67
Browning, Robert, 7, 87, 140, 146
Burdett-Coutts, Georgina, 144
Burne-Jones, Edward, 129, 131
Butt, Dame Clara, 128
Cambridge Philosophical Society, 39, 164
Cannero, Lake Maggiore, 147
Carlyle, Thomas, 54, 90, 115
Carpenter, Edward, 18
Carroll, Lewis, 135
Cayley, Arthur, 2, 22, 29, 160
Chenery, Thomas, 41
Cherry Orchard, The, 8
Chisholm, Roy, ix, 163
Chocorua, 94
Cholmondeley, Mary, 125
Church Times, The, 141
Churchill, Winston, 128

Clare, Lady Castletown, 103
Clebsch, R.F.A., 25
Clifford Algebra, 38, 159, 162, 163
Clifford, Lucy, *A Supreme Moment*, 147
   A Wild Proxy, 80, 147
   A Woman Alone, 82, 149
   About Nellie, 134
   Aunt Anne, 88, 126, 144, 145, 146,
   Chance, The, 155
   Children Busy, Children Glad, Children Naughty, Children Sad, 135
   Dingy House at Kensington, The, 11, 134
   End of her Journey, The, 103, 148
   Eve's Lover, 147.
   Flash of Summer, A, 10, 145
   George Wendern Gave A Party, 126, 155
   Getting Well of Dorothy, The, 151, 152,
   Grey Romance, A, 146
   Hamilton's Second Marriage, 149
   House in Marylebone, The, 153
   Last Touches, The, 149
   Likeness of the Night, 82, 103, 147, 148
   Long Duel, The, 81, 149, 151
   Lost, 87
   Love Letters of a Worldly Woman, 108, 126, 143, 144, 146, 148
   Marie May or Changed Aims, 146
   Marie Zellinger, 146
   Miss Fingal, 74, 125, 153
   Modern Correspondence, A, 108, 127, 143
   Mrs Keith's Crime, 72, 81, 111, 116, 126, 139, 142, 146
   Proposals to Kathleen, 148
   Sir George's Objection, 32, 147, 153
   The New Mother, 136, 137, 138, 139
   Thomas and the Princess, 126
   Wooden Tony, 137, 138, 139
   Woodside Farm, 151
Clifford, Margaret, 56, 72, 73, 74, 77, 81, 86, 95, 112, 113, 119, 121, 124, 126, 127, 129, 137, 151
Clifford, William *Analytical Metrics*, 19
   Applications of Grassman's Extensive Algebra, 38
   Atoms, 33
   Body and Mind, 173

   Common Sense of the Exact Sciences, 2, 39, 42, 124, 159
   Cosmic Emotion, 53, 171, 174
   Elements of Dynamic, 38
   Further note on Biquaternions, 38
   Lectures and Essays, 2, 18, 31, 33, 34, 38, 51, 55, 56
   Mathematical Fragments, 48
   Mathematical Papers, 39, 42
   On the Aims and Instruments of Scientific Thought, 30, 167, 171
   On the Classification of Geometric Algebras, 38
   On the Hypotheses which lie at the Bases of Geometry, 39
   On the Nature of Things in Themselves, 53
   On the Space Theory of Matter, 39
   Preliminary Sketch of Biquaternions, 38
   Seeing and Thinking, 31, 126
   Some of the Conditions of Mental Development, On, 33
   The Analogues of Pascal's Theorem, 17
   The Ethics of Belief, 28, 169
   The First and Last Catastrophe, 33
   The Giant's Shoes, 41
   The Influence upon Morality of a Decline in Religious Belief, 53
   The Philosophy of the Pure Sciences, 33, 163
   The Unseen Universe, 33, 50
   Virchow on the Teaching of Science, 55
Clodd, Edward, 41, 83
Coleridge, Samuel Taylor, 48
Colindale Newspaper Archive, x
Collier, John, 41, 59, 60, 61, 64, 73, 77, 81, 124, 136
Colvin, Sir Sidney, 3, 7, 33, 34, 41, 74, 129, 154
Congress of Liberal Thinkers, 46, 54
Conrad, Joseph, 91
Contemporary Review, The, 34, 39
Conway, Moncure, 29, 30, 41, 42, 54, 56, 154
Corelli, Marie, 81, 134
Cory, William, 61, 63, 65
Cott, Jonathan, 138, 179
Courier, Paul-Louis, 33

Coward, Noel, 80
Craigie, Pearl Mary-Teresa, 124
Crane, Stephen, 91
Crawford, Virginia, 128
Creighton, Mandell, 145
Croom Robertson, G., 29
Cross, John, 70
Daniel Deronda, 54
Darton, Harvey, 136
Darwin, Charles, 16, 18, 31, 34, 83, 144, 168, 170, 172, 173
Debt of Honour, A, 148
Delanghe, R., 162
Demoor, Marysa, ix, 129, 155
Dempwolff, U., 162
Deschamps, Georges, 163
Dialectical Society, The, 34
Dictionary of National Biography, 59, 110, 112
Dilke, Alice, ix
Dilke, Caroline, ix
Dilke, Christopher, ix, 129
Dilke, Ethel, 56, 59, 61, 72, 73, 77, 81, 92, 93, 110, 112, 113, 116, 118, 119, 120, 121, 124, 125, 126, 127, 128, 129, 137, 147, 151, 153
Dilke, Fisher Wentworth, 4, 8, 92, 128, 153
Dilke, Sir Charles, 4, 129, 153
Dirac, P. A.M., 38, 161
Doyle, Sir Arthur Conan, 154
Duckworth, Herbert, 57
Duty of Man, 19
Edel, Leon, 2, 8, 87, 94
Einstein, A, 2, 20, 39, 82, 166
Ekstein, Rudolf, 138
Eliot, George, 1, 2, 6, 43, 44, 45, 46, 54, 69, 70, 71, 72, 119, 124
English Illustrated Magazine, 127
English Literary Studies, 3
Eton Boating Song, 62
Euphorion, 124
Faraday, Michael, 168
Farwell, Ruth, ix, 163, 164, 166, 172
Fawcett, Henry, 25, 29, 112
Femina-Vie Heureuse, 80
Fitzmaurice, Lord George Edmund, 3, 7, 129, 130
Flint, Kate, 149
Ford Madox Ford (Ford Madox Hueffer), 91, 125

Forest Lovers, The, 79
Fortnightly Review, 38, 43, 150
Foster, Professor Michael, 31
Freud, Sigmund, 138
Frost, The Reverend Percival, 22
Fry, Roger, 118
Fueter, R., 162
Gaspey, Thomas, 9, 11
Gavin's Wife, 155
General Theory of Relativity, 39, 164
Gibbs, J. Willard, 2, 160
Gissing, George, 13
Gladstone, William Ewart, 4, 22, 35, 66, 145, 169
Goethe, J.W., 29
Gosse, Edmund, 3, 7, 91, 98, 126
Graham, Kenneth, 123
Grassmann, H.G., 2
Great Exhibition, 1851, 13, 15
Great Men and their Environment, 85
Gregory, Lady, 7
Grote Club, The, 22
Guy Domville, 89
Haldane, R.B., 116, 130
Hamilton, William Rowan, 28, 54, 159, 160, 161, 181
Hardy, Thomas, 4, 41, 80, 112, 140, 142
Harrison, Frederic, 3, 72, 87, 128, 130, 149
Harrison, Mary St Leger, 124
Hartwell, David G., 138
Heine, Heinrich, 36
Helmholtz, Herman von, 2, 171
Helmstadter, R.J., 168
Heroes and Heroines in Children's Stories, 136,
Hestenes, D., 161, 163
Hewlett, Maurice, 79
Highgate Cemetery, 4, 65, 124
Hirst, Thomas, 28
Hodgson, Shadworth, 63
Holmes, Oliver Wendell Jr., 3, 7, 102, 103, 104, 105
Holyoake, G.J., 55
Hooker, J., 29, 168
Houghton, Lord, 20, 33, 140
Howells, William Dean, 96
Humphry Ward, Mrs, 3, 72, 116, 118, 125, 134, 153, 155
Hunt, Violet, 72, 91, 125
Hutton, R.H., 33, 46

Huxley, Marion, 59
Huxley, Thomas, 2, 16, 18, 29, 35, 47, 48, 53, 54, 55, 56, 59, 64, 65, 66, 71, 81, 83, 85, 87, 110, 123, 146, 167
Illustrated London News, 10, 89, 90
Inglis, John (Mrs W.K. Clifford), 126, 155
Institute for Advanced Religious Thought, 30
Interpretation of Dreams, The, 138
Ivory's Theorem, 19
James, Henry, ix, 2, 3, 6, 7, 8, 72, 80, 85, 87, 88, 89, 90, 91, 92, 94, 95, 96, 97, 98, 100, 103, 110, 115, 123, 124, 125, 126, 127, 132, 138, 145, 150, 151, 155
James, Mrs William, 100
James, William, 85, 86, 87, 91, 94
Jowett, Benjamin, 54
Kant, Immanuel, 164, 164, 170, 173
Kendal, Madge, 149
Kennet, Lord, 129
Ker, W.P., 154
King Albert's Book, 128
King's College, London, 16, 18, 28
Kingdon, Fanny (Mary-Anne Bodley), 13
Kinglsey, Charles, 171
Kipling, Rudyard, 3, 7, 72, 73, 74, 76, 77, 78, 79, 91, 108, 118, 127, 128, 156
Klein, F., 2
Knee, Christopher, 164, 166, 172
Knights of Malta, 9
Knowles, James, 34, 52, 145
Ladies' Athenaeum, 88, 103
Lady Love's Journey, 128
Lamb House, 88, 90, 91, 92, 94
Landor, Walter Savage, 13
Lane, John Brandford, 9
Lankester, Sir Ray, 130
Lee, Vernon, 123
Leighton, Lord Frederick, 87
Letters of Rudyard Kipling, The, 74
Lewes, G.H., 1, 43, 44, 45, 46, 50, 69, 103
Light that Failed, The, 75
Lightman, B., 167, 169,
Linton, Eliza Lynn, 41, 46, 130, 134
Literary Fund, The, 9, 11, 71
Little People, The, 41
Lodge, Oliver, 2
London Mathematical Society, 28, 31, 34
Lowell, James Russell, 3, 6, 64, 73, 102, 105, 111, 142, 144, 145

Lubbock, Percy, 2, 31, 69, 91, 94, 98, 130
Lurie, Alison, 136, 138
Lutyens, Sir Edward, 127
Lyell, Charles, 168
MacColl, Norman, 71, 124
Mach, Ernst, 166
Macmillan, Sir Frederick, 31, 33, 37, 51, 65, 72, 73, 79, 80, 124, 125, 126, 139
Macmillan's Magazine, 78, 87
Madeira, 57, 64, 65, 66, 69, 70
Magdalen College, 13
Maitland, Frederick, 57, 111, 120
Malet, Lucas (Mary St Leger Harrison), 124, 134
Mallock, William Hurrell, 53, 54
Manning, Cardinal, 35
Mansfield, Katherine, 118
Marconi, Guglielmo, 66
Marshall, Alfred, 22
Martineau, James, 171
Maugham, Somerset, 8
Maurice, F.D., 35
Maxwell, James Clerk, 2, 21, 22, 28, 48, 66, 160, 166, 169
Mazzini, Giuseppe, 19, 124, 136
Metaphysical Society, 32, 34, 39, 44, 63, 64
Meynell, Alice, 125
Middlemarch, 1, 146
Mill on the Floss, The, 1
Mill, John Stuart, 25
Mind, 39
Modern Thought, 66
Morgan, Charles, 3, 79, 80
Morley, John, 2, 33, 38, 50, 79, 139
Morris, Mowbray, 78
Morrison, Cotter, 41
Morte d'Arthur, 66
My Name is Legion, 79
National Portrait Gallery, x, 59, 98
National Review, The, 146
Natural Theology, 168
Nature, 39, 66
New Republic, The, 53, 54
New Statesman,The, 129
New York Review of Science Fiction, 138
Newman, James R., 42, 43
Newton, Isaac, 4, 164, 170
Nineteenth Century British Culture and Society, 136
Nineteenth Century Fiction, 136

Nineteenth Century, The, 34, 39, 52, 119, 130, 147, 150
Noakes, Burgess, 92
North American Review, 96, 105
Novick, Sheldon M., 103
Observer, The, 79
Old Wives' Tale, The, 145
Origin of Species, 18, 168
Other House, The, 90, 100, 151
Oxford Anthology of Victorian Love Stories, 149, 134
Paddington, Mrs, 93
Paley, William, 168
Pall Mall Gazette, The, 128
Pasteur, Louis, 172
Pater, Walter, 54
Pattison, Mark, 41, 124, 135
Pattison, Mrs Mark (Emilia Dilke), 54, 129
Pearson, K., 2, 32, 42, 114, 124
Penrose, Sir Roger, 162
Perry, Thomas Sergeant, 96
Persse, Jocelyn, 7
Pierce, C.S., 2
Pinero, Sir Arthur Wing, 100
Plain Tales from the Hills, 73
Pollock, Lady Juliet, 26, 29, 41, 63, 103, 117
Pollock, Sir Frederick, 2, 7, 18, 19, 20, 22, 32, 33, 43, 47, 48, 50, 51, 55, 56, 62, 64, 65, 67, 70, 71, 72, 74, 77, 81, 102, 106, 110, 124, 175
Pollock, Sir John, 72, 76, 103
Pollock, Walter Herries, 41, 77
Portrait in a Mirror, 80
Portrait of a Lady, 135
Pound, Ezra, 80
Powell, Frederick York, 41
Power, Professor Edwin, 28
Pritchard, Charles, 34
Problems of Life and Mind, 38,
Prothero, Lady (Fanny), 88
Psychoanalytic Journal for Culture and Science and the Arts, A., 138
Quakers, The, 35
Quarterly Journal of Pure and Applied Mathematics, 17, 19
Quarterly Review, The, 88
Quiver, The, 11, 35, 48, 50, 134, 146, 147
Raleigh, Sir Walter, 22
Religion of an Artist, The, 81
Republican Club, The, 25, 34

Richards, Joan, 165, 167
Riemann, G.F.B., 39, 163
Riesz, Marcel, 161, 163
Robert Elsmere, 125
Robins, Elizabeth, 72, 82, 88
Rogerson, Christina, 97
Romanes, G.J., 29
Royal Astronomical Society, 26
Royal Institution, The, 23, 28, 30, 31, 32, 69
Royal Society, 30, 59, 66, 69, 83, 123, 156
Rumford, Count Benjamin Thompson, 32
Ruskin, John, 35, 54, 168
Russell, Bertrand, 2, 42, 172, 173
Russell, Lord Arthur, 47
Sackville-West, Vita, 120
Salmon, G., 22
Sargent, John Singer, 7, 98, 99
Saturday Review, 134
Savile Club, The, 34, 44, 47, 73, 78, 139
Schaffer, Simon, 166
Schreiner, Olive, 81, 113, 124, 134, 155
Scribner, Charles, 96, 98, 135, 146, 147, 151, 154, 155
Shaw, Bernard, 3, 72, 82, 89, 127
Shaw, R., 162
Shooters Hill, 9, 10
Shorter, Clement, 89, 134
Sidgwick, Henry, 20, 22, 35
Silas Marner, 1
Simeon, Charles, 18
Sinclair, May, 125
Sir George Tressady, 125
Sitwell, Mrs, 7
Smith, H.J.S., 2, 32, 37, 159
Something of Myself, 78
Songs of Dreams, 128
Spectator, 46
Spencer, Herbert, 18, 103
Spottiswood, William, 69, 71
St Germain-en-Laye, 81
Standard, The, 6, 71, 81, 123, 125, 151, 153, 154
Stark, Freya, 154
Stephen, Julia, 57, 87, 94, 103, 114, 115, 121
Stephen, Leslie, 2, 18, 21, 29, 32, 33, 55, 56, 57, 59, 64, 70, 81, 87, 96, 102, 103, 105, 106, 110, 111, 112, 113, 114, 115, 117, 119, 123, 124, 125, 151, 168

Stephen, Vanessa, 94
Stevenson, Robert Louis, 34, 41
Stewart, Balfour, 38
The Unseen Universe, 38
Stokes, George Gabriel, 31
Story of an African Farm, The, 113, 124
Strachey, Lytton, 117
Strand Magazine, 149
Sunday Lecture Society, 31, 32
Swinburne, Algernon Charles, 46, 126, 127, 149, 173
Sylvester, J.J., 2, 22, 29, 39, 48, 54, 118
Tait, Peter, 38
Talland House, 81, 106, 107, 110
Temple Bar, 126, 143, 145, 148
Templeton's Academy, 16, 18
Tennyson, Alfred Lord, 34, 46, 72, 74, 128, 144
Terry, Dame Ellen, x, 3
Thackeray, William Makepeace, 44, 87, 115
Theatre and Friendship, 88
Thirty-nine Articles, 21
Thompson, W.H., 67
Thomson, Sir William (Lord Kelvin), 2, 22, 31, 38, 66, 160
Times, The, 6, 41, 69, 117, 129
To The Lighthouse, 110, 120
Trilobites, 23
Trinity College, 1, 17, 18, 19, 22, 23, 33, 67, 129
Tucker, R., 32
Turn of the Screw, The, 138
Turner, F.M., 168
Tyndall, John, 2, 18, 29, 32, 35, 53, 56, 72, 103, 113, 167, 169, 172, 173, 174
University College, London, x, 1, 2, 27, 28, 29, 33, 41, 42, 51, 103, 124
Valehouse Collection, ix, x, 2, 3, 79, 83, 119, 123, 130
Vanity Fair, 130
Venn, John, 22
Victoria, Queen, 13, 108, 127
Victorian Minds in Crisis, 32
Voltaire, 51, 54
Wallace, A.R., 168
Walpole, Hugh, x, 7, 80, 91, 95
Wells, H.G., 3, 89, 118
Westminster Review, The, 128
Wharton, Edith, 88, 91, 96, 143, 155
Whewell, William, 22, 23
Whitman, Walt, 46, 173
Widnall, Samuel Page, 23
Wilberforce, S., 168
Wilde, Oscar, 143
Will to Believe, The, 85
William and Lucy Clifford Research Group, ix
Without Benefit of Clergy, 73, 77
Wolff, Robert Lee, 136
Women of the Athenaeum, 129
Woolf, Leonard, 117
Woolf, Virginia, 3, 4, 96, 110, 111, 116, 117, 118, 119, 120, 121, 135, 152
Woolson, Constance Fennimore, 88
Woolwich Antiquarian Society, 10
X Club, The, 83
Yeats, W.B., 7
Yonge, Charlotte, 134
Young, R.M., 167, 169